ALLE ZEIT WACH 1842

41. Colloquium der Gesellschaft für Biologische Chemie
5.–7. April 1990 in Mosbach/Baden

# The Molecular Basis of Bacterial Metabolism

Edited by
G. Hauska and R. Thauer

With 84 Figures

Springer-Verlag Berlin Heidelberg New York
London Paris Tokyo Hong Kong Barcelona

Professor Dr. Günter Hauska
FB Biologie
Institut für Botanik
Universitätsstraße 31
W-8400 Regensburg, FRG

Professor Dr. Rudolf K. Thauer
Laboratorium für Mikrobiologie
FB Biologie
Philipps-Universität
Karl-von-Frisch-Straße
W-3500 Marburg, FRG

ISBN-13:978-3-642-75971-0 e-ISBN-13:978-3-642-75969-7
DOI: 10.1007/978-3-642-75969-7

Library of Congress Cataloging-in-Publication Data. Gesellschaft für Biologische Chemie. Colloquium (41st: 1990: Mosbach, Baden-Württemberg, Germany) The molecular basis of bacterial metabolism/ Coloquium der Gesellschaft für Biologische Chemie, 5–7, April 1990 in Mosbach/Baden; edited by G.Hauska and R.Thauer.p.cm.Includes index.ISBN-13:978-3-642-75971-0(U.S.)1.Bacterial metabolism-Congresses. I. Hauska, Günter. II. Thauer, Rudolf. III. Title. QR88.G47 1990 589.9'0133–dc20 90-10340



31/3145-543210 – Printed on acid-free paper

# Preface

The present volume contains 17 lectures of the 41st Mosbach Colloquium of the Gesellschaft für Biologische Chemie, held from April 5–7, 1990 on the topic "The Molecular Basis of Bacterial Metabolism". From the beginning it was not the intention of the organizers to present a comprehensive account, but rather to select new, exciting progress on sometimes exotic reactions of specifically bacterial, mainly anaerobic metabolism. Members of our society had contributed to this progress to an extent that greatly stimulated the scientific exchange with international colleagues during the days in Mosbach.

The editors hope that this stimulation will be conveyed to the readers of the articles, which reach from the biochemistry of methanogenesis, via anaerobic radical reactions, metal biochemistry in hydrogen and nitrogen metabolism, conversions of light – and redox energy, to the regulation of metabolic adaptation, and the attempts to bioengineer novel pathways for the degradation of xenobiotica. We believe that the book represents a highly progressive field of overlapping disciplines, comprising microbiology and molecular genetics, chemistry of biomimetic interest, and biophysics, and that it gives insight into the impact modern technologies have on microbiological research today. The colloquium was generously supported by the Deutsche Forschungsgemeinschaft, the Paul-Martini-Stiftung, and the Fonds für Biologische Chemie.

A. Trebst, G. Schäfer, and D. Oesterhelt were a great help in preparing the program and we wish to thank them for their advice. We would also like to express our gratitude to Mrs. M. Schmidt, Mrs. E. Preiß, Mrs. G. Gerling, Dr. E. Truscheit, and Dr. K.T. Beaucamp for sharing the burden of organizing the colloquium.

The effort of Springer-Verlag, especially of Mrs. Andrea Schlitzberger, for rapid publication of this book is gratefully appreciated.

September 1990

G. Hauska, Regensburg
R. Thauer, Marburg

# Contents

# Contributors

You will find the addresses at the beginning of the respective contribution

# Novel Coenzymes of Archaebacteria

R.S. Wolfe[1]

## 1 Introduction

To effect enzyme-catalyzed reactions of small molecules, nature has evolved coenzymes. During the first 60 years of this century, most of the coenzymes of metabolism (diphosphothiamin, ATP, biotin, coenzyme A, pyridoxal phosphate, folic acid, FMN-FAD, NAD-NADP, lipoic acid, cobamides, and quinones) were defined. Most of these coenzymes were found to be cosmopolitan. Their distribution among prokaryotes and eukaryotes strongly supported the doctrine of comparative biochemistry, i.e., what was true for bacteria also was true for plants and animals, there being only one major exception: cobamides apparently played no role in the evolution of plants.

During this era of biochemistry the enzymology of methanogenic bacteria (methanogens) lay unexplored. Experiments were confined to a few laboratories and were limited to the physiology of whole cells. The techniques of microbiology were not adequate for the routine isolation and cultivation of such strict anaerobes as the methanogens. It was not until the mid-1960s that reliable and effective techniques were developed for the mass culture of methanogens. When kilogram quantities of cells became available by mass culture of cells on hydrogen and carbon dioxide, the stage was set for a serious study of the biochemistry of methanogenesis. However, even then progress was slow, for techniques had to be evolved for the fractionation of oxygen-sensitive cell extracts.

This chapter presents a brief historical overview of the six novel coenzymes of methanogenesis, their discovery, and role. Methanogenesis from hydrogen and carbon dioxide, a system common to all but a few methanogens, will be discussed first. Most studies have been carried out with extracts of a thermophilic methanogen, *Methanobacterium thermoautotrophicum*, the enormous advantage being that enzymes could be fractionated at room temperature (40 °C below the optimal growth temperature of the organism).

## 2 Results and Discussion

### 2.1 *Methanofuran*

In the reduction of $CO_2$ to methane, the first coenzyme to be involved is methanofuran; the first stable product of this new method of $CO_2$ fixation is formyl-

[1]Department of Microbiology, 131 Burrill Hall, University of Illinois, 407 S. Goodwin, Urbana, IL 61801, USA

41. Colloquium Mosbach 1990
The Molecular Basis of Bacterial Metabolism

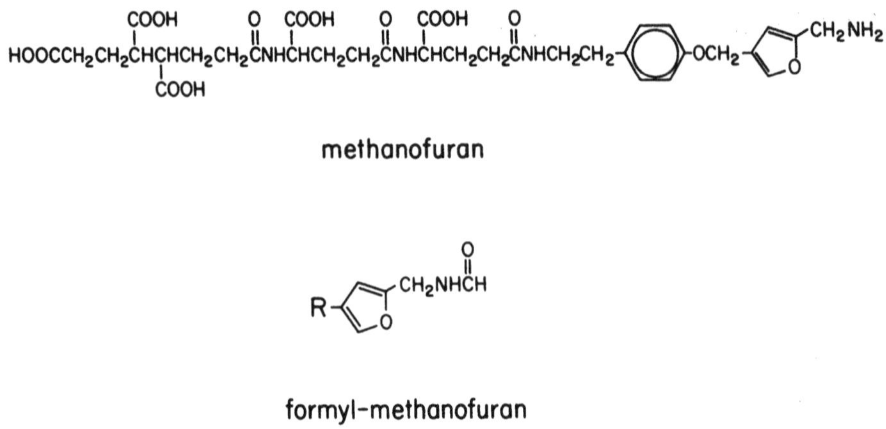


**Fig. 1.** The coenzyme methanofuran and its formylated derivative. The nitrogen atom of the primary amine next to the furan ring becomes the $C_1$ carrier for the first reduced product of $CO_2$ fixation, a formyl group

methanofuran (Fig. 1). A simple experiment led to the discovery of methanofuran. When cell extracts were passed through a small column of Sephadex G-25, the pass-through fraction was not able to catalyze the reduction of $CO_2$ to $CH_4$ (Romesser and Wolfe 1982). Boiled cell extract or eluates from the column contained a factor that restored activity. Using this assay, the factor was purified; its structural formula was established as 4-[N-(4,5,7-tricarboxyheptanoyl-γ-L-glutamyl-γ-L-glutamyl-)-p-(β-aminoethyl) phenoxymethyl]-2-(aminomethyl) furan (Leigh et al. 1984). The nitrogen atom of the primary amine attached to the furan ring is the carrier for the first $CO_2$ reduction step (Leigh et al. 1985). The mechanism of $CO_2$ reduction is unknown, but the reaction has been followed in reverse (Leigh et al. 1985), and the oxidation of formylmethanofuran to $Co_2$ coupled to reduction of a dye has been studied (Börner et al. 1989).

## 2.2 *Methanopterin*

One of the fluorescent factors which was readily observed during column chromatography of cell extracts was factor 342 ($F_{342}$), the compound being named for its strong absorbance peak at 342 nm (Gunsalus and Wolfe 1978). It was identified as a pterin (Keltjens et al. 1983) and named methanopterin. The complete structure of methenyltetrahydromethanopterin was determined (van Beelan et al. 1984). The structure of tetrahydromethanopterin ($H_4MPT$) is shown in Fig. 2A. A factor required for formaldehyde conversion to methane, the formaldehyde activation factor (FAF), was found to be tetrahydromethanopterin (MW 776); its oxidized state was found to be methanopterin, MW 772 (Escalante-Semerena et al. 1984a). The reaction product with formaldehyde was identified as methylene-tetrahydromethanopterin (Fig. 2D) which could be oxidized to methenyl-tetrahydromethanopterin (Fig. 2C) or reduced with borohydrided to 5-methyl-tetrahydromethanopterin (Fig. 2E; Escalante-Semerena et al.

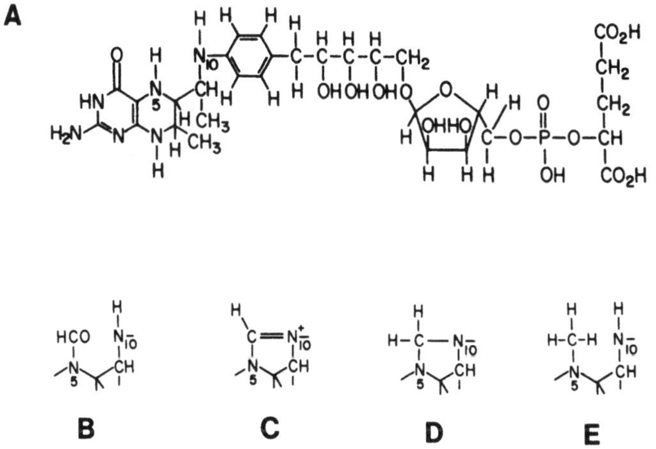

**Fig. 2A-E.** The coenzyme methanopterin and its $C_1$ derivatives. **A** Tetrahydromethanopterin ($H_4MPT$); **B** 5-formyl-$H_4MPT$; **C** 5-10 methenyl-$H_4MPT$; **D** 5-10 methylene $H_4MPT$; **E** 5-methyl-$H_4MPT$

1984b). The 5-formyl-tetrahydromethanopterin derivative (Fig. 2B) was identified as a stable compound, whereas the 10-formyl derivative was unstable (Donnelly et al. 1985); properties of these formyl derivatives of tetrahydromethanopterin stand in contrast to similar derivaties of the tetrahydrofolic acid. Although methanopterin is a more complex molecule than folic acid, the pterin moieties are similar except for methylation at C-7 and C-11, and the $C_1$ derivaties of the two compounds are similar, involving the $N^5$ and $N^{10}$ positions of the pterin.

### *2.3 Coenzyme $F_{420}$*

Cells of methanogens exhibit a bright blue-green fluorescence under UV light in an expifluorescence microscope. This fluorescence is due to high concentrations of an unusual flavin derivative named $F_{420}$ for its strong absorption at 420 nm. $F_{420}$ was first observed during column chromatography (Cheeseman et al. 1972), and its structure (Fig. 3) was determined to be the N-(N-L-lactyl-γ-L-glutamyl)-L-glutamic acid phosphodiester of 7,8-didemethyl-8-hydroxy-5-deazariboflavin 5′ phosphate (Eirich et al. 1978). The structure was confirmed by synthesis (Ashton et al. 1979). Because of the carbon atom at position 5, the reduced compound (Fig. 3) cannot act as a normal semiquinone in electron transport; it is a two-electron carrier that is involved in hydride transfer reactions. A recent review summarizes the chemistry of $F_{420}$ (Walsh 1986). When $F_{420}$ is reduced, the strong absorption at 420 nm disappears, and the compound becomes colorless and nonfluorescent.

### *2.4 Coenzyme M*

The first novel coenzyme for which there was an assay was discovered by following the formation of methane from methylcobalamin by cell extracts. A factor present

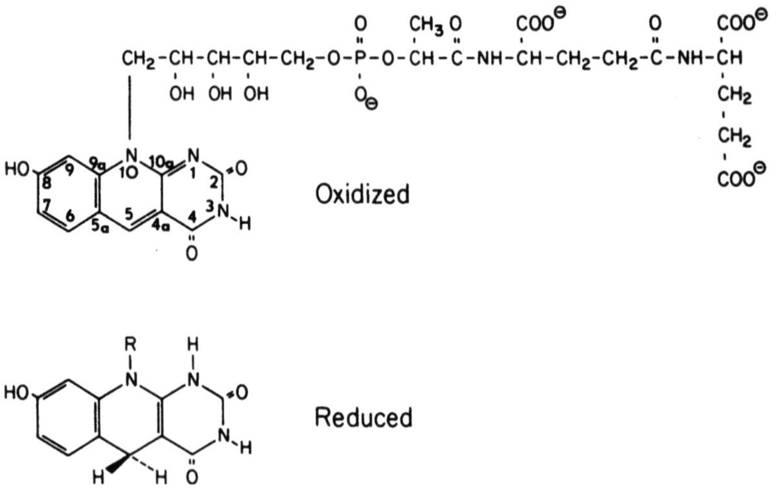


**Fig. 3.** Coenzyme $F_{420}$, a 7,8-didemethyl-8-hydroxy-5-dezariboflavin derivative

in fractionated extracts was required for the conversion of the methyl group of methylcobalamin to methane. The name "coenzyme M" was given to this compound because it was involved in some way in methyl transfer (McBride and Wolfe 1971). Coenzyme M (Fig. 4A) was determined to be 2-mecaptoethanesulfonic acid (HS-CoM) and was found to carry a methyl group (Fig. 4B) as 2-(methylthio)ethanesulfonic acid ($CH_3$-S-CoM; Taylor and Wolfe 1974a). An extensive survey of tissue samples from the biological world with an assay that would detect 10 pmol revealed that coenzyme M is found only in methanogens (Balch and Wolfe 1979).

### *2.5 $F_{430}$, a Nickel-Containing Tetrahydrocorphin*

$F_{430}$ was first observed during column chromatography of cell extracts as a nonfluorescent yellow band by Jean LeGall. The compound was named $F_{430}$ because of its strong absorption at 430 nm (Gunsalus and Wolfe 1978). Independently in two laboratories, $F_{430}$ was found to contain nickel (Whitman and Wolfe 1980; Diekert et al. 1980a). Evidence was obtained for the incorporation of $\delta$-aminolevulinic acid into $F_{430}$ (Diekert et al. 1980b) suggesting that the compound was a tetrapyrrole. The structure of $F_{430}$ (Fig. 5) was obtained in Eschenmoeser's laboratory, where it was shown to be a nickel-containing tetrahydrocorphin (Pfaltz et al. 1982; Livingston et al. 1984). The pentaacid form of $F_{430}$ is now believed to be the native, active form of the coenzyme. Recently, by use of two-dimensional $^1H$ and $^{13}C$ NMR studies, the complete assignments have been

A $HS-CH_2-CH_2-SO_3^-$

B $CH_3-S-CH_2-CH_2-SO_3^-$

**Fig. 4A,B.** Coenzyme M. **A** 2-mercaptoethanesulfonic acid (HS-CoM); **B** 2-(methylthio)ethanesulfonic acid ($CH_3$-S-CoM)

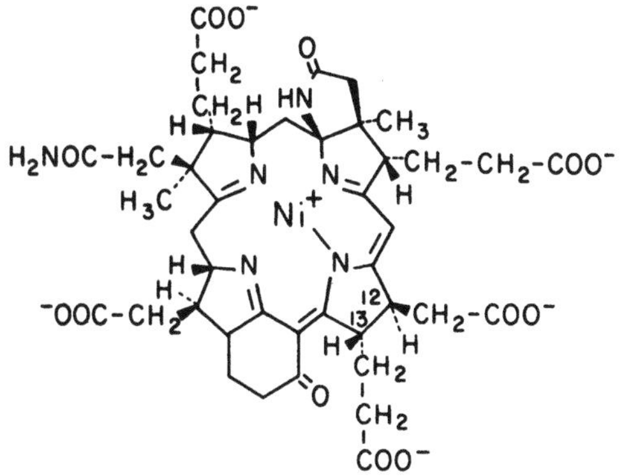


Factor $F_{430}$

**Fig. 5.** The tetrapyrrole Factor 430 ($F_{430}$). The native pentacid form is shown. The novel tetrahydrocorphin nucleus binds nickel

made for the protons and carbon atoms of native $F_{430}$ in the laboratory of Summers (Won et al. 1990). So far, $F_{430}$ has been found only in methanogens.

## *2.6 Component B, 7-mercaptoheptanoylthreonine Phosphate (HS-HTP)*

In the late 1970s Gunsalus discovered and developed an assay for the heat-stable, dialyzable factor that was required for methane formation from $CH_3$-S-CoM (Gunsalus and Wolfe 1980). The factor was named component B, two crude protein fractions A and C also being required to effect methanogenesis. Component B proved to be unusually difficult to purify with a huge loss of activity occurring during fractionation. Finally the compound was shown to contain a thiol, and its structure (Fig. 6) was elucidated as N-(7-mercaptoheptanoyl)threonine-$O^3$-phosphate, (HS-HTP; Noll et al. 1986) and confirmed by chemical synthesis (Noll et al. 1987). When the 7-carbon chain was increased to 8 or decreased to 6, the derivatives became inhibitors of methane formation (Ellermann et al. 1988).

## *2.7 Role of the Six Coenzymes in the Reduction of $CO_2$ to $CH_4$*

Nature's strategy in the reduction of $CO_2$ to $CH_4$ is to pass the $C_1$ group attached to a coenzyme as it is sequentially reduced. In Fig. 7, the enzyme reactions are arranged to emphasize that in *Methanobacterium thermoautotrophicum*, a product of the terminal reaction is involved in the activation and reduction of $CO_2$ to the formyl level.

The first stable product of $CO_2$ fixation is formyl-methanofuran (Fig. 7, reaction 1). The mechanism of this reaction is unknown, but it has been shown that addition of the heterodisulfide (CoM-S-S-HTP) stimulates the formation of for-

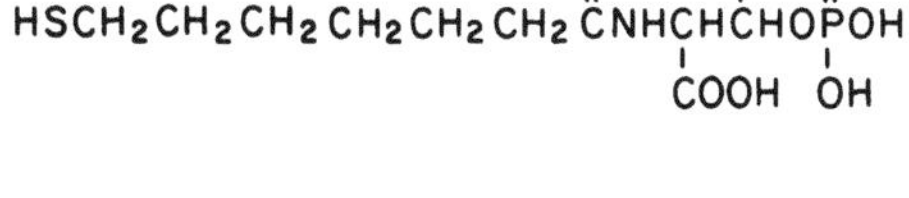


**Fig. 6.** 7-mercaptoheptanoylthreonine phosphate (HS-HTP), component B of the methylreductase system

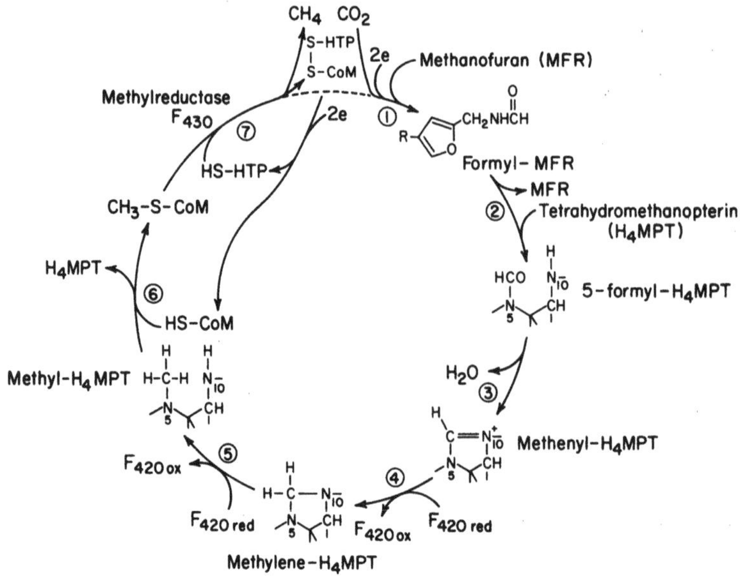


**Fig. 7.** Proposed pathway for the reduction of $CO_2$ to $CH_4$ in *Methanobacterium thermoautotrophicum*. Only the active sites of the coenzymes methanofuran and tetrahydromethanopterin which bind a $C_1$ group are shown (after the Journal of Biological Chemistry)

mylmethanofuran 42 fold (Bobik and Wolfe 1988). The reverse reaction, catalyzed by formylmethanofuran dehydrogenase, has yielded to fractionation by use of an assay in which methyl viologen serves as the electron acceptor (Börner et al. 1989). In reaction 2 (Fig. 7), the formyl group is transferred by the enzyme formylmethanofuran:tetrahydromethanopterin formyltransferase, the product of this reaction being the 5-formyl derivative of $H_4$MPT (Donnelly et al. 1985). This enzyme has been cloned and sequenced, and an enzyme that is active at thermophilic temperatures has been expressed in *E. coli* (DiMarco et al. 1990). In reaction 3 (Fig. 7), water is removed from the formyl group and a methenyl derivative of $H_4$MPT is formed by the enzyme, 5-10-methenyltetrahydromethanopterin cyclohydroloase. This enzyme has been purified to homogeneity (DiMarco et al. 1986). The methenyl group of methenyl-$H_4$MPT is next reduced to a methylene group (reaction 4, Fig. 7) by the enzyme, methylenetetrahydromethanopterin:coenzyme $F_{420}$ oxidoreductase, the electrons being provided by the reduce deazaflavin coenzyme $F_{420}$ (Hartzell et al. 1985). Reduction of the methylene group to methyl-$H_4$MPT was first performed by use of borohydride (Escalante-Semerena et al. 1984a). Fractionation of the enzyme which carried out this reduction (reaction 5, Fig. 7) proved to be difficult. However, a breakthrough was recently made; the enzyme has been purified to homogeneity and reduced $F_{420}$ was identified as the electron donor (Brömmelstroet et al. 1990). Therefore, reaction 5 is catalyzed by methyltetrahydromethanopterin:coenzyme $F_{420}$ oxidoreductase. Reaction 6 (Fig. 7) most likely involves two enzyme steps in analogy with methyl-group transfer in *Methanosarcina barkerii*, where a cobamide enzyme accepts the methyl group from methanol (van der Meijden et al. 1983a,b) and a noncobamide methyltransferase (Taylor and Wolfe 1974b) transfers the methyl group from the cobamide to HS-CoM to form $CH_3$-S-CoM.

In reaction 7 (Fig. 7), $CH_3$-S-CoM is the substrate for the methylreductase where the methyl group is reduced to $CH_4$. When component C of the methylreductase system was purified to homogeneity, it was found to be yellow with an absorption maximum at 421 nm. Analysis of the protein yielded an $M_r$ of 300 000, with three subunits in an $\alpha_2\beta_2\gamma_2$ configuration with apparent molecular weights of 68 000, 45 000 and 38 500 (Ellefson and Wolfe 1981; Ellefson et al. 1982) and a stoichiometry of 2 mol of $F_{430}$ per mol of protein. An active form of component C ($C_a$) was found to catalyze the reduction of $CH_3$-S-CoM to $CH_4$ in a reaction mixture with an artificial electron donor, indicating that component C is the methylreductase enzyme (Ankel-Fuchs et al. 1987). Later HS-HTP was found to be the electron donor for the methylreductase (Bobik et al. 1987; Ellerman et al. 1988), and the products were $CH_4$ and the heterodisulfide of coenzyme M and HS-HTP (CoM-S-S-HTP). $CH_3$-S-CoM + HS-HTP $\rightarrow$ $CH_4$ + CoM-S-S-HTP.

Fraction A of the methylreductase system from *Methanobacterium thermoautotrophicuim* strain ΔH (Gunsalus and Wolfe 1980) has been resolved into protein fractions A1-A4, of which only A2 has been purified to homogeneity. These results have been summarized (Rouviere and Wolfe 1988). The function of these fractions is open to speculation; A2 and A3 may be involved in conversion of component $C_i$ to $C_a$, and A1 may provide the electrons for the reduction of CoM-S-S-HTP. Evidence suggests that the conversion of $C_i$ to $C_a$ involves the reduction of the nickel atom of $F_{430}$ frm $Ni^{II}$ to $Ni^{I}$ (Albracht et al. 1988). The characteristic signals of MCR-red 1 are believed to represent the first stage of reduction, and in the next stage $Ni^{I}$ is believed to react with HS-HTP, producing a ligand (MCR-red 2) in which $Ni^{I}$ coordination is radically changed. It is this species which is believed to react with $CH_3$-S-CoM to produce $CH_4$.

## *2.8 Methanogenesis from Methanol*

All methanogens employ a methyl group as their terminal electron acceptor. This methyl group may be the product of $CO_2$ reduction as in *Methanobacterium* (Fig. 7), or the methyl group may be a portion of a substrate molecule such as methanol. To grow on methanol as the sole substrate, *Methanosarcina* oxidizes 1 mol of methanol completely to $CO_2$ and employs the methyl groups of 3 mol of methanol as electron acceptors. A proposed model for this system (Börner et al. 1989; Fischer and Thauer 1989; Möller-Zinkhan et al. 1989) is shown in Fig. 8, where at the top left the methyl group of methanol is activated by a corrinoid protein and transferred to form methyl-$H_4$MPT. The scheme shown in Fig. 7 is operated in reverse, as methyl-$H_4$MPT is oxidized to methylene-$H_4$MPT, which is then oxidized to methenyl-$H_4$MPT. A cyclohydrolase adds water to form 5-formyl-$H_4$MPT, and the formyl group is transferred to methanofuran. Formyl-methanofuran dehydrogenase oxidizes the formyl group to $CO_2$. At each oxidative step, two electrons are removed to electron carriers. Coenzyme $F_{420}$ may serve as the carrier from the oxidation of methyl and methylene-$H_4$MPT, and from formylmethanofuran. The heterodisulfide CoM-S-S-HTP serves as the electron sink for the electron carrier,

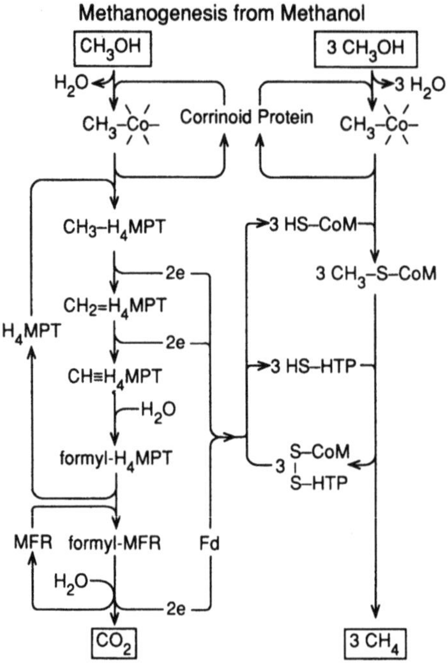


**Fig. 8.** Proposed model for the formation of methane from methanol by *Methanosarcina*. On the *left* one molecule of methanol is oxidized completely to $CO_2$, generating 3 electron pairs. Each pair is transferred to the heterodisulfide reductase on the *right* where HS-HTP serves as the electron donor for methyl group reduction via the methylreductase

as the heterodisulfide reductase produces HS-CoM and HS-HTP for the methylreductase reaction. Each methyl group for the methylreductase reaction is derived from methanol, as shown at the top right of Fig. 8. A corrinoid protein and a methyltransferase (van der Meijden et al. 1983a,b) are involved in activation and transfer of the methyl group of methanol to form $CH_3$-S-CoM.

## 2.9 *Methanogenesis from Acetate*

It was not until the mid-1970s that a reliable method for growing *Methanosarcina* on acetate was developed (Smith and Mah 1978). Since that time results from a number of investigations (Lovley et al. 1984; Nelson and Ferry 1984; Eikmanns and Thauer 1984; Krzycki et al. 1985; Laufer et al. 1987; Jablonski et al. 1990) suggest that acetate may be converted to methane by a model pathway similar to that shown in Fig. 9. Acetate is activated by a conventional acetate kinase to form acetyl-phosphate, and acetyl-CoA is formed by phosphotransacetylase (Lundie and Ferry 1989). Acetyl-CoA becomes the substrate for a five-subunit enzyme complex that contains a dimeric nickel-iron containing carbon monoxide dehydrogenase and a dimeric iron-sulfur corrinoid protein. Carbon-2 of the acetyl group binds to the cobalt atom of the corrinoid protein, and carbon-1 of the acetyl group binds to the dimeric Ni-Fe protein; during catalysis the carbon-cobalt bond is cleaved and the methyl group is transferred to $H_4$MPT by a methyl transferase and then to HS-CoM to form $CH_3$-S-CoM. The carbon monoxide group is oxidized by the Fe-Ni carbon monoxide dehydrogenase dimeric protein with ferredoxin as the electron acceptor. Electrons are then donated to the heterodisulfide reductase

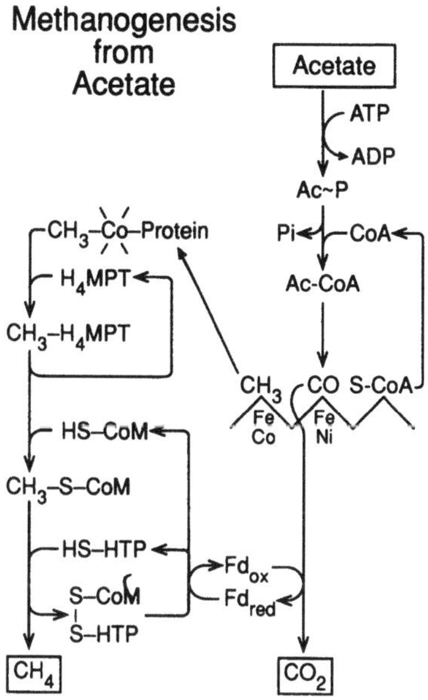


**Fig. 9.** Proposed model for methanogenesis from acetate by *Methanosarcina*. Carbon monoxide dehydrogenase provides the electrons for the heterodisulfide reductase producing HS-CoM and HS-HTP for the methylreductase

where CoM-S-S-HTP is reduced, generating HS-CoM and HS-HTP for participation in the $CH_3$-S-CoM methylreductase reaction.

## 3 Coenzymes in *Archaeoglobus*

The extreme thermophilic archaebacterium, *Archaeoglobus fulgidus*, grows under anaerobic conditions by oxidizing lactate to $CO_2$ with sulfate being reduced to $H_2S$ (Stetter 1988; Zellner et al. 1989). Recently this pathway of lactate oxidation (Fig. 10) has been shown to involve the methyl, methylene, methenyl, and formyl derivatives of $H_4MPT$, as well as formylmethanofuran when the methyl group of lactate is oxidized to $CO_2$ (Möller-Zinkhan et al. 1989). Although *Archaeoglobus* is not a methanogenic archaebacterium, it possesses the unique coenzymes of methanogenes, $H_4MPT$ and methanofuran, which function as $C_1$ carriers in the reverse direction to that shown in Fig. 7 for $CO_2$ reduction. In addition, *Archaeoglobus* possess coenzyme $F_{420}$. The pathway shown in Fig. 10 has been proposed by Möller-Zinkhan et al. (1989) with the bold arrows indicating reactions carried out by the membrane fraction of broken cells, whereas the thin arrows indicate reactions carried out by enzymes in the soluble fraction. Although coenzyme $F_{420}$, $H_4MPT$, and methanofuran have been found in achaebacteria other than methanogens, $F_{430}$, coenzyme M, and HS-HTP appear to be associated only with methanogens and the methylreductase reaction.

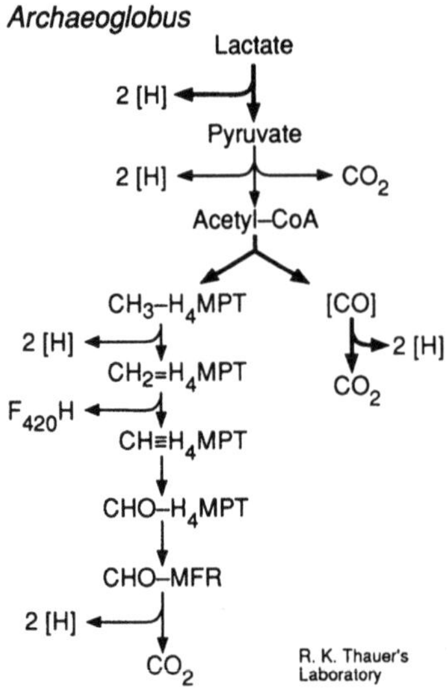


**Fig. 10.** Methyl group conversion to $CO_2$ by *Archaeoglobus*, an archaebacterium that employs the coenzymes tetrahydromethanopterin and methanofuran in an oxidative $C_1$ pathway of methyl group oxidation to carbon dioxide. (Archives for Microbiology)

*Acknowledgments.* Research from the author's laboratory was supported by research grants from the National Institutes of Health, National Science Foundation, and Department of Energy.

## References

Albracht, S.P.J., Ankel-Fuchs, D., Böcher, R., Ellerman, J., Moll, J., van der Zwaan, J.W. & Thauer, R.K. (1988) Five new EPR signals assigned to nickel-coenzyme M reductase from *Methanobacterium thermoautotrophicum*, strain Marburg. Biochem. Biophys. Acta 955:86–102

Ankel-Fuchs, D., Böcher, R., Thauer, R.K., Noll, K.M. & Wolfe, R.S. (1987) 7-mercaptoheptanoylthreonine phosphate functions as component B in ATP-independent methane formation from methyl-CoM with reduced cobalamin as electron donor. FEBS Lett. 213:123–127

Ashton, W.T., Brown, R.D., Jacobson, F. & Walsh, C. (1979) Synthesis of 7,8-didemethyl-8-hydroxy-5-deazariboflavin and confirmation of its identity with the deazaisoalloxazine chromophore of *Methanobacterium* redox coenzyme $F_{420}$. J. Am. Chem. Soc. 101:4419–4420

Balch, W.E. & Wolfe, R.S. (1979) Specificity and biological distribution of coenzyme M (2-mercaptoethanesulfonic acid). J. Bacteriol. 137:256–263

Bobik, T.A. & Wolfe, R.S. (1988) Physiological importance of the heterodisulfide of coenzyme M and 7-mercaptoheptanoylthreonine phosphate in the reduction of carbon dioxide to methane in *Methanobacterium*. Proc. Natl. Acad. Sci. USA 85:60–63

Bobik, T.A., Olson, K.D., Noll, K.M. & Wolfe, R.S. (1987) Evidence that the heterodisulfide of coenzyme M and 7-mercaptoheptanoylthreonine phosphate is a product of the methylreductase reaction in *Methanobacterium*. Biochem. Biophys. Res. Commun. 149:455–460

Börner, G., Karrasch, M. & Thauer, R.K. (1989) Formylmethanofuran dehydrogenase activity in cell extracts of *Methanobacterium thermoautotrophicum* and of *Methanosarcina barkeri*. FEBS Letters 244:21–25

Brömmelstroet, te, B.W., Hensgens, C.M.H., Keltjens, J.T., van der Drift, C. & Vogels, G.D. (1990) Purification and properties of 5–10-methylene-tetrahydromethanopterin reductase, a coenzyme $F_{420}$-dependent enzyme from *Methanobacterium thermoautotrophicum* strain ΔH. J. Biol. Chem. 265:1852–1857

Cheeseman, P., Toms-Wood, A. & Wolfe, R.S. (1972) Isolation and properties of a fluorescent compound, $factor_{420}$, from *Methanobacterium* strain M.o.H. J. Bacteriol. 112:527–531

Diekert, G., Klee, B. & Thauer, R.K. (1980a) Nickel, a component of factor $F_{430}$ from *Methanobacterium thermoautotrophicum*. Arch. Microbiol. 124:103-106

Diekert, G., Jaenchen, R. & Thauer, R.K. (1980b) Biosynthetic evidence for a nickel tetrapyrrole structure of factor $F_{430}$ from *Methanobacterium thermoautotrophicum*. FEBS Lett. 119:118-120

DiMarco, A.A., Donnelly, M.I. & Wolfe, R.S. (1986) Purification and properties of the 5,10-methenyltetrahydromethanopterin cyclohydrolase from *Methanobacterium thermoautotrophicum*. J. Bacteriol. 168:1372-1377

DiMarco, A.A., Sment, K.A. Konisky, J. & Wolfe, R.S. (1990) The formylmethanofuran:tetrahydromethanopterin formyltransferase from *Methanobacterium thermoautotrophicum*: nucleotide sequence and functional expression of the cloned gene. J. Biol. Chem. 265:472-476

Donnelly, M.I., Escalante-Semerena, J.C., Rinehart, K.L., Jr. & Wolfe, R.S. (1985) Methenyl-tetrahydromethanopterin cyclohydrolase in cell extracts of *Methanobacterium*. Arch. Biochem. Biophys. 242:430-439

Eikmanns, B. & Thauer, R.K. (1984) Catalysis of an isotopic exchange between $CO_2$ and the carboxyl group of acetate by *Methanosarcina barkeri* grown on acetate. Arch. Microbiol. 138:365-370

Eirich, L.D., Vogels, G.D. & Wolfe, R.S. (1978) Proposed structure for coenzyme $F_{420}$ from *Methanobacterium*. Biochemistry 17:4583-4593

Ellefson, W.L. & Wolfe, R.S. (1981) Component C of the methylreductase system of *Methanobacterium*. J. Biol. Chem. 256:4259-4262

Ellefson, W.L., Whitman, W.B. & Wolfe, R.S. (1982) Nickel-containing factor $F_{430}$: chromophore of the methylreductase of *Methanobacterium*. Proc. Natl. Acad. Sci. USA 79:3707-3710

Ellermann, J., Hedderich, R., Böcher, R. & Thauer, R.K. (1988) The final step in methane formation. Eur. J. Biochem. 172:669-677

Escalante-Semerena, J.C., Leigh, J.A., Rinehart, K.L. Jr. & Wolfe, R.S. (1984a) Formaldehyde activation factor, tetrahydromethanopterin, a coenzyme of methanogenesis. Proc. Natl. Acad. Sci. USA 81:1976-1980

Escalante-Semerena, J.C., Rinehart, K.L. Jr. & Wolfe, R.S. (1984b) Tetrahydromethanopterin, a carbon carrier in methanogenesis. J. Biol. Chem. 259:9447-9455

Fischer, R. & Thauer, R.K. (1989) Methyltetrahydromethanopterin as an intermediate in methanogenesis from acetate in *Methanosarcina barkeri*. Arch. Microbiol. 151:459-465

Gunsalus, R.P. & Wolfe, R.S. (1978) Chromophoric factors $F_{342}$ and $F_{430}$ of *Methanobacterium*. FEMS Microbiol. Lett. 3:191-193

Gunsalus, R.P. & Wolfe, R.S. (1980) Methyl coenzyme M reductase from *Methanobacterium thermoautotrophicum*: resolution and properties of the components. J. Biol. Chem. 255:1891-1895

Hartzell, P.L., Zvilius, G., Escalante-Semerena, J.C. & Donnelly, M.I. (1985) Coenzyme $F_{420}$ dependence of the methylenetetrahydromethanopterin dehydrogenase of *Methanobacterium thermoautotrophicum*. Biochem. Biophys. Res. Commun. 133:884-890

Jablonski, P.E., DiMarco, A.A., Bobik, T.A., Cabell, M.C. & Ferry, J.G. (1990) Protein content and enzyme activities in methanol and acetate-grown *Methanosarcina thermophila*. J. Bacteriol. 172:1271-1275

Keltjens, J.T., Huberts, M.J., Laarhoven, W.H. & Vogels, G.D. (1983) Structural elements of methanopterin, a novel pterin present in *Methanobacterium thermoautotrophicum*. Eur. Biochem. 130:537-544

Krzycki, J.A., Lehman, L.J. & Zeikus, J.G. (1985) Acetate catabolism by *Methanosarcina barkeri*: evidence for involvement of carbon monoxide dehydrogenase, methyl coenzyme M, and methylreductases. J. Bacteriol. 163:1000-1006

Laufer, K., Eikmanns, B., Frimmer, V. & Thauer, R.K. (1987) Methanogenesis from acetate by *Methanosarcina barkeri*: catalysis of acetate formation from methyliodide, $CO_2$, and $H_2$ by the enzyme system involved. Z. Naturforsch. 42c:360-372

Leigh, J.A., Rinehart, K.L. Jr. & Wolfe, R.S. (1984) Structure of methanofuran, the carbon dioxide reduction factor of *Methanobacterium thermoautotrophicum*. J. Am. Chem. Soc. 106:3636-3640

Leigh, J.A., Rinehart, K.L. Jr. & Wolfe, R.S. (1985) Methanofuran (carbon dioxide reduction factor), a formyl carrier in methane production from carbon dioxide in *Methanobacterium*. Biochemistry 24:995-999

Livingston, D.A., Pfaltz, A., Schreiber, J., Eschenmoser, A., Ankel-Fuchs, D., Moll, J., Jaenchen, R. & Thauer, R.K. (1984) Zur Kenntnis des Faktors $F_{430}$ aus methanogenen Bakterien: Struktur des proteinfreien Faktors. Helv. Chim. Acta 67:334–351
Thauer, R.K. (1984) Zur Kenntnis des Faktors $F_{430}$ aus methanogenen Bakterien: Struktur des proteinfreien Faktors. Helv. Chim. Acta 67:334–351
Lovley, D.R., White, R.H. & Ferry, J.G. (1984) Identification of methyl coenzyme M as an intermediate in methanogenesis from acetate in *Methanosarcina*. J. Bacteriol. 160:521–525

Lundie, L.L. & Ferry, J.G. (1989) Activation of acetate by *Methanosarcina thermophila*: purification and characterization of phosphotransacetylase. J. Biol. Chem. 264:18392–18396

McBride, B.C. & Wolfe, R.S. (1971) A new coenzyme of methyl transfer, coenzyme M. Biochemistry 10:2317–2324

Möller-Zinkhan, D., Börner, G. & Thauer, R.K. (1989) Function of methanofuran, tetrahydromethanopterin, and coenzyme $F_{420}$ in *Archaeoglobus fulgidus*. Arch. Microbiol. 152:362–368

Nelson, M.J.K. & Ferry, J.G. (1984) Carbon monoxide dependent methyl coenzyme M methylreductase in acetotrophic *Methanosarcina*. J. Bacteriol. 160:526–532

Noll, K.M., Rinehart, K.L. Jr., Tanner, R.S. & Wolfe, R.S. (1986) Structure of component B (7-mercaptoheptanoylthreonine phosphate) of the methylcoenzyme M methylreductase system of *Methanobacterium thermoautotrophicum*. Proc. Natl. Acad. Sci. USA 83:4238–4242

Noll, K.M., Donnelly, M.I. & Wolfe, R.S. (1987) Synthesis of 7-mercaptoheptanoylthreonine phosphate and its activity in the methylcoenzyme M methylreductase system. J. Biol. Chem 262:513–515

Pfaltz, A., Jaun, B., Fässler, A., Eschenmoser, A., Jaenchen, R., Gilles, H.H., Diekert, G. & Thauer, R.K. (1982) Zur Kenntnis des Faktors $F_{430}$ aus methanogenen Bakterien: Struktur des porphinoiden Ligand Systems. Helv. Chim. Acta. 65:828–865

Romesser, J.A. & Wolfe, R.S. (1982) CDR factor, a new coenzyme required for carbon dioxide reduction to methane by extracts of *Methanobacterium thermoautotrophicum*. Zentralbl, Bakteriol. Hyg. I. Abt. Orig. C3, 271–276

Rouviére, P.E. & Wolfe, R.S. (1988) Novel biochemistry of methanogenesis. J. Biol. Chem. 263:7913–7916

Smith, M.R. & Mah, R.A. (1978) Growth and methanogenesis by *Methanosarcina* strain 227 on acetate and methanol. Appl. Environ Microbiol. 36:870–879

Stetter, K.O. (1988) *Archaeoglobus fulgidus* gen. nov., sp. nov.: a new taxon of extremely thermophilic archaebacteria. Syst. Appl. Microbiol. 10:172–173

Taylor, C.D. & Wolfe, R.S. (1974a) Structure and methylation of coenzyme M ($HSCH_2CH_2SO_3^-$). J. Biol. Chem. 249:4879–4885

Taylor, C.D. & Wolfe, R.S. (1974b) A simplified assay for coenzyme M ($HSCH_2CH_2SO_3^-$): resolution of methylcobalamin-coenzyme M methyltransferase and use of sodium borohydride. J. Biol. Chem. 249:4886–4890

van Beelen, P., van Neck, J.W., de Cock, R.M., Vogels, G.D., Guijt, W. & Haasnoot, C.A.G. (1984) 5-10-methenyl-5,6,7,8-tetrahydromethanopterin, a one-carbon carrier in the process of methanogenesis. Biochemistry 23:4448–4454

van der Meijden, P., Heijthuysen, H.J., Pouwels, A., Houven, F., van der Drift, C. & Vogels, G.D. (1983a) Methyltransferases involved in methanol conversion by *Methanosarcina barkeri*. Arch. Microbiol. 134:238–242

van der Meijden, P., Jansen, L.P.J.M., van der Drift, C. & Vogels, G.D. (1983b) Involvement of corrinoids in the methylation of coenzyme M (2-mercaptoethanesulfonic acid) by methanol and coenzymes from *Methanosarcina barkeri*. FEMS Microbiol. Lett. 19:247–251

Walsh, C. (1986) Naturally occurring 5-deazaflavin coenzymes: biological redox roles. Acc. Chem. Res. 19:216–221

Whitman, W.B. & Wolfe, R.S. (1980) Presence of nickel in factor $F_{430}$ from *Methanobacterium bryantii*. Biochem. Biophys. Res. Commun. 92:1196–1201

Won, H., Summers, M.F., Olson, K. & Wolfe, R.S. (1990) Two-dimensional NMR studies of native coenzyme $F_{430}$. J. Am. Chem. Soc. 112:2178–2184

Zellner, G., Stackebrandt, E., Kneifel, H., Messner, P., Sleytr, U.B., Conway de Macrio, E., Zabel, H.P., Stetter, K.O. & Winter, J. (1989) Isolation and characterization of a thermophilic sulfate reducing archaebacterium, *Archaeoglobus fulgidus* strain Z. Syst. Appl. Microbiol. 11:151–160

# Alternatives to the Calvin Cycle and the Krebs Cycle in Anaerobic Bacteria: Pathways with Carbonylation Chemistry

G. FUCHS[1]

## 1 Introduction

Anaerobic microorganisms differ from aerobic organisms in their pathways of central carbon metabolism. A variety of them undergo a chemosynthesis, i.e., they are enabled to grow autotrophically with $CO_2$ as sole source of cell carbon. Others grow heterotrophically by oxidizing organic compounds completely to $CO_2$. However, the Calvin cycle for $CO_2$ fixation does not operate in any of these anaerobic autotrophs, and in many of the heterotrophs the Krebs cycle is lacking.

Acetyl-coenzyme A is the central intermediate in anaerobic carbon metabolism. It can be formed in a noncyclic reductive pathway from two molecules of $CO_2$, which are separately reduced to enzyme-bound carbonyl and pterin-bound methyl groups; these $C_1$-units are linked together to activated acetic acid. Vice versa, acetyl-CoA can be cleaved directly to a carbonyl and a methyl group which are oxidized separately via a similar noncyclic route to two molecules of $CO_2$. Hence, both $CO_2$ fixation and acetyl-CoA oxidation are catalyzed by a similar set of enzymes. These pathways, based on carbonylation chemistry, will be described and discussed. Since most of the literature has been reviewed elsewhere (Wood 1985; Wood et al. 1986a,b; Ljungdahl 1986; Fuchs 1986, 1989; Thauer et al. 1977, 1989; Thauer 1988), only some selected, original papers will be quoted.

## 2 Alternative(s) to the Calvin Cycle

Different alternatives to the Calvin cycle have been found in the following major groups of strictly anaerobic autotrophic bacteria. These organisms obtain energy by electron transport phosphorylation coupled to anaerobic respiration or to photosynthetic light reaction.

1. The methanogenic Archaebacteria, which gain energy from the reduction of $CO_2$ or $C_1$-compounds to $CH_4$, or from acetate disproportionation to $CO_2$ and $CH_4$;
2. The acetogenic bacteria, which gain energy by reducing $CO_2$ or $C_1$-compounds to acetate;
3. The sulfate-reducing (sulfidogenic) bacteria which gain energy by reducing oxidized sulfur compounds to $H_2S$;
4. The phototrophic green sulfur bacteria;

[1]Abteilung Angewandte Mikrobiologie, Universität Ulm, W-7900 Ulm, FRG

41. Colloquium Mosbach 1990
The Molecular Basis of Bacterial Metabolism

5. The thermophilic gliding phototrophic bacteria;
6. The thermoacidophilic, sulfur-dependent Archaebacteria which gain energy by reducing or oxidizing sulfur to $H_2S$ and sulfate, respectively; finally,
7. In aerobic thermophilic Knallgasbacteria which may be phylogenetically closely related to strict anaerobes; they gain energy from the Knallgas reaction.

Even the reduction of oxidized metals may provide energy for autotrophic growth.
Three alternative $CO_2$ fixation mechanisms have been recognized (Fig. 1):

1. The reductive acetyl CoA pathway in the groups 1, 2, 3 (see below);
2. The reductive citric acid cycle in the groups 3, 4, 6, 7 (Evans et al. 1966; Ivanovsky et al. 1980; Kandler and Stetter 1981; Shiba et al. 1985; Schäfer et al. 1989; for review, see Fuchs 1989)
3. A yet to be defined new pathway in group 5 (Holo and Sirevåg 1986; Holo 1989).

Acetate synthesis from $CO_2$ was studied first in acetogenic bacteria by H.A. Barker (1956), H.G. Wood, L. Ljungdahl, and their associates (for a historical account, see Wood 1985). It was found that autotrophic $CO_2$ fixation in methanogenic bacteria (and in many other autotrophs) follows a similar principle, the reductive acetyl CoA/CO dehydrogenase pathway (Knight et al. 1966; Taylor et al. 1976; Zeikus et al. 1977; Daniels and Zeikus 1978; Fuchs et al. 1978; Fig. 2). This noncyclic pathway is characterized by de novo synthesis of acetyl CoA from 2 $CO_2 + 8[H]$ and has a complex metallo-organic biochemistry. Especially noteworthy is the role of an enzyme-bound carbon monoxide in enzyme catalysis; metal carbonyls have not been encountered before in the biochemistry of carbon metabolism.

Acetyl CoA synthesis requires two reductive $CO_2$ fixations: One $CO_2$-fixing enzyme is formate dehydrogenase, a Mo- or W, Fe/S, (Se) enzyme reducing $CO_2$ to formate; this reaction is then followed by ATP-dependent N-formyltetrahydropterin formation and the reduction of formyl- to methyltetrahydropterin (Ljungdahl 1986). The second, most unusual $CO_2$-fixing enzyme is carbon monoxide dehydrogenase, a rather complex Ni, Fe/S, Zn enzyme, which reduces

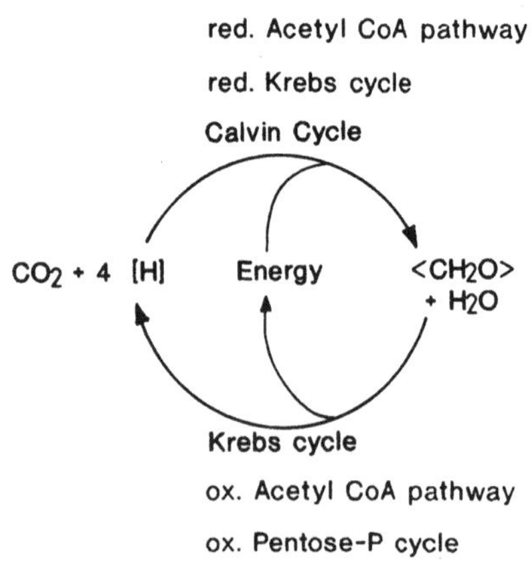


**Fig. 1.** Some alternatives to the Calvin cycle and the Krebs cycle in carbon metabolism of anaerobic bacteria

CO dehydrogenase/Acetyl CoA synthase Complex

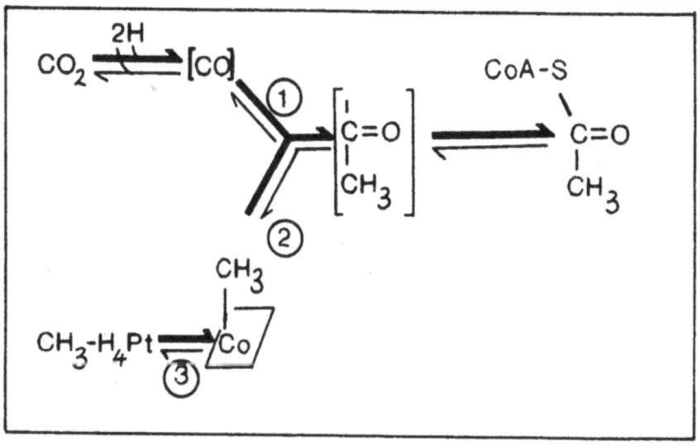


| | | | |
|---|---|---|---|
| 1 CO dehydrogenase | 2 Ni, 1 Zn, 11 Fe, 14 S | | |
| 150 kDa $\alpha\beta$ | $[4Fe\text{-}4S]^{2+/1+}$ | -430 mV | |
| $\alpha$ 70 kDa | 2Fe | -260 mV | |
| $\beta$ 80 kDa | | -530 mV | |
| 2 *Corrinoid protein* ($CH_3$ carrier) | | | |
| 88 kDa $\alpha\beta$ | 1 Corrinoid base off : | | |
| $\alpha$ 34 kDa | $[CO]^{3+/2+/1+}$ | | +300 mV/-500 mV |
| $\beta$ 55 kDa | $[4Fe\text{-}4S]^{2+/1+}$ | | -510 mV |
| 3 *Methyltransferase* | | | |
| 59 kDa $\alpha_2$ | | | |
| $\alpha$ 30 kDa | | | |

**Fig. 2.** Essential reactions of acetyl CoA synthesis (→) or cleavage (←) both catalyzed by the CO dehydrogenase/acetyl CoA synthase enzyme complex. The $CH_3$-tetrahydropterin ($CH_3$-$H_4$Pt) is being formed reductively from $CO_2$ + 6 reducing equivalents (H) during autotrophic $CO_2$ fixation, or oxidized during heterotrophic acetyl CoA oxidation. The scheme also summarizes the composition of the enzyme complex and the midpoint potential (pH 7) of the redoxactive centers. /Co/ represents the corrinoid protein; [ ] indicates enzyme-bound intermediates

$CO_2$ to an enzyme (Ni)-bound carbonyl. This enzyme complex also brings about the combination of the methyl, the carbonyl, and coenzyme A to acetyl CoA. The transfer of the methyl from methyltetrahydropterin to CO dehydrogenase requires a corrinoid protein as methyl carrier and a methyltransferase. These enzymes are probably organized in a complex (Fig. 3). $CO_2$ reduction to the carbonyl level is driven by proton motive force (Diekert et al. 1986) and CO dehydrogenase therefore is supposed to be somehow membrane-associated. The product of the reaction sequence, acetyl CoA, is not assimilated via the glyoxylate bypass, but is reductively carboxylated to pyruvate catalyzed by pyruvate synthase (for pyruvate synthase, see Kerscher and Oesterheldt 1982).

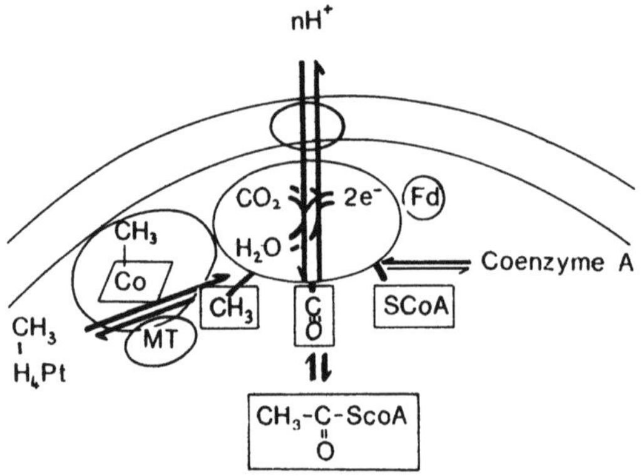


**Fig. 3.** Tentative scheme illustrating the interaction of CO dehydrogenase/acetyl CoA synthase with several proteins in acetyl Co synthesis/cleavage and the role of membranes and proton motive force. *Fd* Ferredoxin; *MT* methyl transferase; *$H_4Pt$* tetrahydropterin; Co corrinoid protein

## 3 Biochemical and Energetic Aspects of Carbonylation

Many of the biochemical and most of the energetic aspects of this reaction sequence are barely understood. Several metals (Fe, Co, Ni, Zn, Mo, W) and trace elements (Se), one-carbon carrying coenzymes, and prosthetic groups (tetrahydropterins, corrinoids), as well as electron carriers (flavins, deazaflavins, iron-sulfur proteins, quinones, pyridine nucleotides) are involved. The most intriguing questions concern the role of metals and metallo-organic complexes in the course of the reactions. The spectroscopic and redox properties of the metal proteins and the active sites are being studied intensively (Pezacka and Wood 1988; Harder et al. 1989; Lindahl et al. 1990a,b; Lu et al..1990). So far, the oxygen-sensitive CO dehydrogenase/acetyl CoA synthase complex has not been prepared in a homogeneous, biologically active form.

The bacterial acetyl CoA synthesis may be considered a biological analogue of the industrial synthesis of acetate from $CO + CH_3OH$ in the Monsanto process or of the Fischer-Tropsch synthesis of hydrocarbons from water gas. These processes involve organometallic chemistry. The specific role of membranes, membrane proteins, and membrane-bound electron carriers in this 8-electron redox reaction has been documented, but it is only now becoming clear how the endergonic reduction of $CO_2$ to the carbonyl and formaldehyde level may be coupled to subsequent exergonic reactions. These reactions may be linked by a proton and a sodium motive force (Blaut and Gottschalk 1985; Schönheit and Beimborn 1985). The overall process must require much less than the energetic equivalent of 1 mol of ATP per mol acetyl CoA formed from 2 $CO_2$.

## 4. An Alternative to the Krebs Cycle

Several strict anaerobic bacteria are able to oxidize organic matter via acetyl CoA completely to $CO_2$ (Widdel and Pfennig 1977; for literature, see Thauer et al. 1989). The electron acceptors may be, e.g., sulfate, oxidized metals, or even $H_2$-consuming symbiontic bacteria. In some of them the citric acid cycle with various, most interesting, and ingenious modifications has been found (Thauer 1988). It appears, however, that the majority of the anaerobic acetate oxidizers oxidize acetyl CoA to $CO_2$ via an oxidative acetyl CoA/CO dehydrogenase pathway (Schauder et al. 1986; Lee and Zinder 1988; Thauer et al. 1989). This is in principle a reversal of the reductive $CO_2$ fixation mechanism in autotrophs. Consequently, acetyl CoA is cleaved by the CO dehydrogenase complex into a metal-bound carbonyl, a methyl corrinoid, and CoA. The carbonyl is oxidized to $CO_2$ directly by the CO dehydrogenase enzyme, and a proton motive force is built up. The methyl is transferred to a tetrahydropterin and is then oxidized via formate to $CO_2$. It is rather unlikely, although as yet unproven, that acetate oxidation to $CO_2$ uses exactly the same set of enzymes. One energetic difference between the oxidative citric acid cycle and this novel pathway is the differing midpoint potential of the oxidoreductases involved (Fig. 4).

## 5 Autotrophy Versus Heterotrophy

Two question are especially intriguing in those facultatively autotrophic/heterotrophic bacteria which can grow either autotrophically ($CO_2 + H_2$) or heterotrophically (e.g., lactate + sulfate), and which use the same type of reactions in the reductive and the oxidative direction: (1) Do the enzymes of the

| Pathway | Redox couple | $E^{o'}$ (mV) | average $E^{o'}$ (mV) |
|---|---|---|---|
| | 2 $CO_2$/acetate | -291 | |
| Citric acid cycle | 2-Oxoglutarate + $CO_2$/isocitrate | -364 | -247 |
| | Succinyl-CoA + $CO_2$/2-oxoglutarate | -491 | |
| | Fumarate/succinate | + 32 | |
| | Oxaloacetate/malate | -166 | |
| Acetyl CoA/CO dehydrogenase pathway | $CH_2{=}FH_4/CH_3{-}FH_4$ | -117 | -339 |
| | $CH{\equiv}FH_4/CH_2{=}FH_4$ | -295 | |
| | $CO_2/HCOO^-$ | -421 | |
| | $CO_2$/CO | -524 | |

**Fig. 4.** Redox couples, estimated midpoint potentials of individual redox couples, and estimated average midpoint potential of the four oxidoreductase reactions of the oxidative citric acid cycle and the oxidative acetyl CoA/CO dehydrogenase pathway (Thauer 1988). $CH_2 = FH_4$ = methylenetetrahydrofolate; $CH_3 - FH_4$ = methyltetrahydrofolate; $CH \equiv FH_4$ = methenyltetrahydrofolate

autotrophic and the heterotrophic pathways differ and (2) how is the carbon and electron flux directed in a reaction sequence which seems to be close to equilibrium. There are indications for two different CO dehydrogenases depending on whether $CO_2$ has to be reduced to a carbonyl (to form acetyl CoA), or whether the carbonyl (obtained by cleaving acetyl CoA) has to be oxidized (Terlesky et al. 1986). The genes for the CO dehydrogenase/acetyl CoA synthase complex are organized in a cluster (Roberts et al. 1989). It is expected that this genetic approach will soon clarify whether there is only one or more CO dehydrogenase gene clusters or operons in these bacteria.

## 6 Variations of the Principle

This novel pathway in anaerobic bacteria exhibits a remarkable biochemical variability. Formate can be a free intermediate; or, in Archaebacteria, it is linked as a carbamate and N-formyl group to another new coenzyme, methanofuran, before being transferred to tetrahydropterin (Thauer et al. 1989; see also Wolfe, this Vol.). Virtually all coenzymes, electron carriers (which are not all known), and prosthetic groups involved vary. Consequently, the enzymes must differ to some degree. These biochemical variations are partly due to the evolutionary distance of the anaerobes, and partly to the different functions of these reactions in different organisms.

This set of $C_1$- and $C_2$-metabolizing enzymes can be used for many different purposes, in addition to those mentioned before, provided that additional enzymes and coenzymes are available. They all involve carbonyl chemistry (Fig. 5):

1. $CO_2$ reduction to acetyl CoA (cell carbon fixation);
2. Acetyl CoA oxidation to 2 $CO_2$ (energy generation);

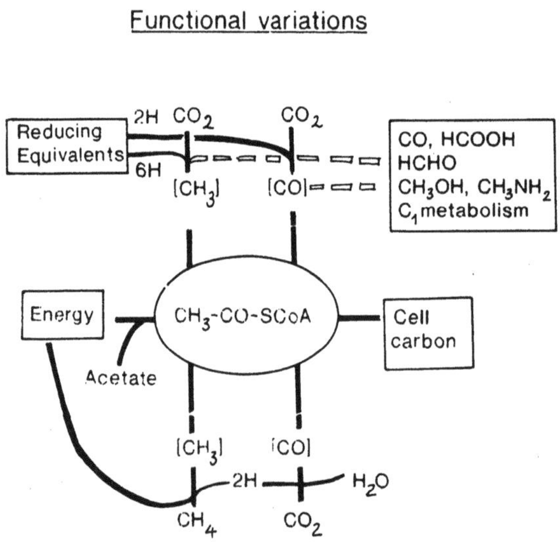


**Fig. 5.** Different metabolic functions of the reactions and enzymes of the acetyl CoA/CO dehydrogenase pathway and the role of carbonyl biochemistry in different anaerobic bacteria

3. Acetate diproportionation to $CO_2$ and $CH_4$ (energy generation);
4. Assimilation of $C_1$ compounds such as CO, HCOOH, $CH_3OH$, and $CH_3NH_2$ into acetyl CoA (cell carbon fixation);
5. Oxidation of such $C_1$ compounds to $CO_2$, or disproportionation to $CO_2$ plus $CH_4$ or acetate (energy generation);
6. $CO_2$ reduction to acetate (energy generation).

## References


Barker, H.A. (1956) Bacterial fermentations. Wiley, New York

Blaut, M. & Gottschalk, G. (1985) Evidence for a chemiosmotic mechanism of ATP synthesis in methanogenic bacteria. Trends Biochem. Sci. 10:486–489

Daniels, L. & Zeikus, J.G. (1978) One-carbon metabolism in methanogenic bacteria: analysis of short-term fixation products of $^{14}CO_2$ and $^{14}CH_3OH$ incorporated into whole cells. J. Bacteriol. 136:75–84

Diekert, G., Schrader, E. & Harder, W. (1986) Energetics of CO formation and CO oxidation in cell suspensions of *Acetobacterium woodii*. Arch. Microbiol. 144:386–392

Evans, M.C.W., Buchanan, B.B. & Arnon, D.I. (1966) A new ferredoxin-dependent carbon reduction cycle in a photosynthetic bacterium. Proc. Natl. Acad. Sci. USA 55:928–934

Fuchs, G. (1986) $CO_2$ fixation in acetogenic bacteria: variations on a theme. FEMS Microbiol. Rev. 39:181–213

Fuchs, G. (1989) Alternative pathways of autotrophic $CO_2$ fixation. In: Schlegel, H.G. & Bowien, B. (eds) Autotrophic bacteria. Science Tech. Publishers, Madison and Springer, Berlin Heidelberg New York, pp 365–382

Fuchs, G., Stupperich, E. & Thauer, R.K. (1978) Acetate assimilation and the synthesis of alanine, aspartate and glutamate in *Methanobacterium thermoautotrophicum*. Arch. Microbiol. 117:61–66

Harder, S.R., Lu Wi-P., Feinberg, B.A. & Ragsdale, S.W. (1989) Spectroelectrochemical studies of the corrinoid/iron sulfur protein involved in acetyl coenzyme A synthesis by *Clostoridum thermoaceticum*. Biochemistry 28:9080–9087

Holo, H. (1989) *Chloroflexus aurantiacus* secretes 3-hydroxypropionate, a possible intermediate in the assimilation of $CO_2$ and acetate. Arch. Microbiol. 151:252–256

Holo, H. & Sirevåg, R. (1986) Autotrophic growth and $CO_2$ fixation of *Chloroflexus aurantiacus*. Arch. Microbiol. 145:173–180

Ivanovsky, R.N., Sintsov, N.V. & Kondratieva, E.N. (1980) ATP-linked citrate lyase activity in the green sulfur bacterium *Chlorobium limicola* forma *thiosulfatophilum*. Arch. Microbiol. 128:239–241

Kandler, O.K. & Stetter, K.O. (1981) Evidence for autotrophic $CO_2$ assimilation in *Sulfolobus brierleyi* via a reductive carboxylic acid pathway. Zbl. Bakt. Hyg., I. Abt. Orig. C 2:111–121

Kerscher, L. & Oesterhelt, D. (1982) Pyruvate: ferredoxin oxidoreductase – new findings on an ancient enzyme. Trends Biochem. Sci. 7:371–374

Knight, M., Wolfe, R.S. & Elsden, S.R. (1966) The synthesis of amino acids by *Methanobacterium omelianskii*. Biochem. J. 99:76–86

Lee, M.J. & Zinder, S.H. (1988) Carbon monoxide pathway enzyme activities in a thermophilic anaerobic bacterium grown acetogenically and in a synthrophic acetate-oxidizing coculture. Arch. Microbiol. 150:513–518

Lindahl, P.A., Münck, E. & Ragsdale, S.W. (1990a) CO dehydrogenase from *Clostridium thermoaceticum*. J. Biol. Chem. 265:3873–3879

Lindahl, P.A., Ragsdale, S.W. & Münck, E. (1990b) Mössbauer study of CO dehydrogenase from *Clostridium thermoaceticum*. J. Biol. Chem. 265:3880–3888

Ljungdahl, L.G. (1986) The autotrophic pathway of acetate synthesis in acetogenic bacteria. Ann. Rev. Microbiol. 40:415–450

Lu, W.P., Harder, S.R. & Ragsdale, S.W. (1990) Controlled potential enzymology of methyl transfer reactions involved in acetyl-CoA synthesis by CO dehydrogenase and the corrinoid/iron-sulfur protein from *Clostridium thermoaceticum*. J. Biol. Chem. 265:3124–3133

Pezacka, E. & Wood, H.G. (1988) Acetyl-CoA pathway of autotrophic growth. J. Biol. Chem. 263:16000–16006

Roberts, D.L., James-Hagstrom, J.E., Garvin, D.K., Gorst, C.M., Runquist, J.A., Bauer, J.R., Haase, F.C. & Ragsdale, S.W. (1989) Cloning and expression of the gene cluster encoding key proteins involved in acetyl-CoA synthesis in *Clostridum thermoacetiucum*: CO dehydrogenase, the corrinoid/Fe-S protein, and methyltransferase. Proc. Natl. Acad. Sci. USA 86:32–36

Schäfer, S., Götz, M., Eisenreich, W., Bacher, A. & Fuchs, G. (1989) $^{13}C$-NMR study of autotrophic $CO_2$ fixation in *Thermoproteus neutrophilus*. Eur. J. Biochem. 184:151–156

Schauder, R., Eikmanns, B., Thauer, R.K., Widdel, F. & Fuchs, G. (1986) Acetate oxidation to $CO_2$ in anaerobic bacteria via a novel pathway not involving reactions of the citric acid cycle. Arch. Microbiol. 145:162–172

Schönheit, P. & Beimborn, D.B. (1985) Presence of a $Na^+/H^+$ antiporter in *Methanobacterium thermoautotrophicum* and its role in $Na^+$ dependent methanogenesis. Arch. Microbiol. 142:354–361

Shiba, H., Kawasumi, T. Igarashi, Y., Dokoma, T. & Minoda Y. (1985) The $CO_2$ assimilation via the reductive tricarboxylic acid cycle in an obligately autotrophic aerobic hydrogen-oxidizing bacterium, *Hydrogenobacter thermophilus*. Arch. Microbiol. 142:198–203

Taylor, G.T. Kelly, D.P. & Pirt, S.J. (1976) Intermediary metabolism in methanogenic bacteria (*Methanobacterium*). In: Schlegel, H.G., Gottschalk, G. & Pfennig, N. (eds) Microbial production and utilization of gases ($H_2$, $CH_4$, CO). Goltze Verlag, Göttingen, pp 173–180

Terlesky, K.C., Nelson, M.J.K. & Ferry, J.G. (1986) Isolation of an enzyme complex with carbon monoxide dehydrogenase activity containing corrinoid and nickel from acetate-grown *Methanosarcina thermophila*. J. Bacteriol. 168:1053–1058

Thauer, R.K. (1988) Citric-acid cycle, 50 years on. Modifications and an alternative pathway in anaerobic bacteria. Eur. J. Biochem. 176:497–508

Thauer, R.K., Jungermann, K. & Decker, K. (1977) Energy conservation in chemotrophic anaerobic bacteria. Bact. Rev. 41:100–180

Thauer, R.K., Möller-Zinkhan, D. & Spormann, A.M. (1989) Biochemistry of acetate catabolism in anaerobic chemotrophic bacteria. Annu. Rev. Microbiol. 43:43–67

Widdel, F. & Pfennig, N, (1977) A new anaerobic, sporing, acetate-oxidizing bacterium, *Desulfotomaculum* (emend.) *acetoxidans*. Arch. Microbiol. 112:119–122

Wood, H.G. (1985) Then and now. Ann. Rev. Biochem. 54:1–41

Wood, H.G. Ragsdale, S.W. & Pezacka, E. (1986a) The acetyl-CoA pathway of autotrophic growth. FEMS Microbiol. Rev. 39:345–362

Wood, H.G. Ragsdale, S.W. & Pezacka, E. (1986b) The acetyl-CoA pathway: a newly discovered pathway of autotrophic growth. TIBS 11:14–18

Zeikus, J.G., Fuchs, G., Kenealy, W. & Thauer, R.K. (1977) Oxidoreductases involved in cell carbon synthesis of *Methanobacterium thermoautotrophicum*. J. Bacteriol. 132:604–613

# Amino Acid Fermentation: Coenzyme $B_{12}$-Dependent and -Independent Pathways

W. BUCKEL[1]

## 1 Introduction

The fermentation of amino acids to ammonia, carbon dioxide, and short-chain fatty acids is an important step in the mineralization of proteins in soil, sewage sludge, and marine and fresh water sediments. In addition, amino acids are anaerobically degraded in the intestine and in other parts of humans and animals which are badly supplied with oxygen. The majority of the organisms which are involved in these fermentations are Clostridia and other gram-positive bacteria. Only a few representatives are found among the Proteobacteria and in the Bacteroides group (for reviews see Barker 1961, 1981).

Usually the fermentation pathways are redox reactions in which amino acids act as electron donators as well as acceptors. In the classical Stickland reaction in *Clostridium sporogenes* (Stickland 1934), alanine, leucine, or valine are oxidized, whereas glycine, proline, or tryptophan are reduced. Later it was shown that one amino acid alone may cover both functions, as did glutamate in *C. tetanomorphum* (Barker 1937) or in *Micrococcus aerogenes* (*Peptostreptococcus assacharolyticus*; Whiteley 1957), alanine in *C. propionicum*, and glycine in *Diplococcus glycinophilus* (*P. micros*; Cardon and Barker 1946). Today the Stickland reaction appears to be a speciality of some Clostridia, whereas the ability to ferment single amino acids is more common.

The first steps of the oxidations are very similar to those in aerobic organisms. Many amino acids are converted to the corresponding 2-oxoacids by oxidative desamination, transamination, or $\alpha$, $\beta$-elimination, followed by oxidative decarboxylation to the corresponding fatty acid. The reductions, however, are a speciality of the anaerobes. The main chemical problem thereby appears to be the removal of the $\alpha$-amino group, since there are five different pathways known. The key reactions are the following:

1. Reductive cleavage of the $C_\alpha$-N bond;
2. $\alpha$, $\beta$-Elimination of ammonia;
3. Carbon-carbon rearrangement of glutamate to 3-methylaspartate;
4. Shift of the $\alpha$-amino group to the $\beta$-position;
5. Conversion to (*R*)-2-hydroxyacyl-CoA and dehydration to enoyl-CoA.

This chapter will show the role of these pathways in the fermentation of the proteinogenes amino acids and discuss the mechanisms of these unusual reactions.

[1] Laboratorium für Mikrobiologie, Fachbereich Biologie, Philipps-Universität, W-3550 Marburg, FRG

41. Colloquium Mosbach 1990
The Molecular Basis of Bacterial Metabolism

Finally, a sixth section will deal with the fermentation of such omega-amino acids as 4-aminobutyrate and 5-aminovalerate.

## 2 Results and Discussion

### *2.1 Reductive Cleavage of the $C_\alpha$-N Bond*

Some Clostridia and Peptostreptococci, as well as *Eubacterium acidaminophilum* are able to reduce glycine to $NH_4^+$ and acetate. Clostridia also catalyze the analogous reduction of proline to 5-aminovalerate (Stickland 1934; Barker 1961, 1981; Zindel et al. 1988). The reducing agent in vivo is probably NADH in combination with an electron transport chain. In vitro and with purified enzymes, a dithiol such as dithioerythritol (DTE) is usually applied. Tanaka and Stadtman (1979) showed that the reduction of 1 mol glycine was coupled to the synthesis of 1 mol ATP from ADP and inorganic phosphate. This unusual result was recently confirmed, with the only difference being that acetyl phosphate was detected as primary product instead of ATP [Eq. (1)]. The formation of ATP, as observed earlier, was due to the presence of acetate kinase in the enzyme preparation (Arkowitz and Abeles 1989).

$$\text{Glycin} + DTE_{red} + P_i + H^+ = \text{acetyl phosphate} + NH_4^+ + DTE_{ox} + H_2O;$$
$$\Delta G^{\circ\prime} = -16.8\ \text{kJ/mol} \quad (1)$$

The glycine reductase complex from *C. sticklandii* is composed of three proteins, A, B and C, from which only A was purified to homogeneity. It was characterized as a selenocysteine containing glycoprotein (m = 12 kDa; Cone et al. 1976). Besides the selenoprotein, no other cofactors are required for the reduction of glycine. Protein B probably contains pyruvoyl residues as prosthetic groups (Tanaka and Stadtman 1979) which could facilitate the release of $NH_4^+$ (Arkowitz and Abeles 1989). (*R*) and (*S*)-[2-$^2$H, 2-$^3$H]glycine was reduced to (*R*) and (*S*)-acetate, respectively. The reduction occurred stereospecifically with inversion (Barnard and Akhtar 1979), therefore a carboxymethylene radical as intermediate appears very unlikely (see Sect. 3). However, Arkowitz and Abeles (1990) found recently that the selenocysteine residue was carboxymethylated by glycine during the reaction. Thus the initial step of glycine reduction is apparently the nucleophilic displacement of the amino group of glycine by the selenol residue. Subsequent reductive elimination could yield ketene which acetylates a thiol group on protein C. Final cleavage of this thiolester by inorganic phosphate affords the product acetyl phosphate (Fig. 1).

In the enzymatic reduction of D-proline, free 5-aminovaleric acid and not the acyl phosphate is formed [Eq. (2); Barker 1981].

$$\text{D-Proline} + DTE_{red} = \text{5-aminovalerate} + DTE_{ox}; \quad (2)$$

It may be possible that a substrate chain phosphorylation is thermodynamically not feasible in this reaction. In any case, the reduction of glycine is more favorable with regard to the entropy since two products, acetylphosphate and $NH_4^+$ are generated. The reduction of proline in *C. sticklandii* is catalyzed by a single enzyme. It is a

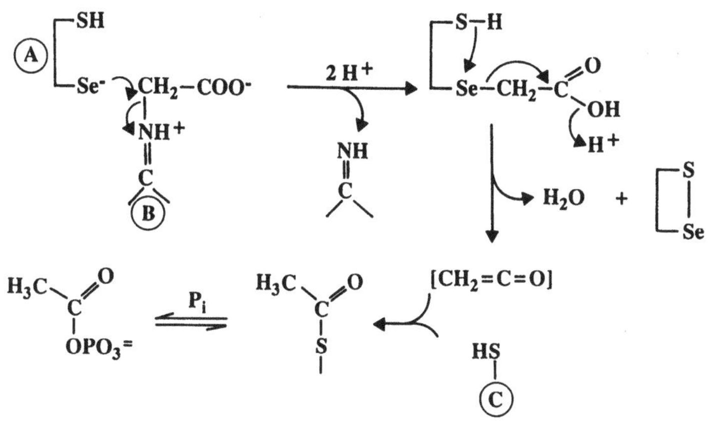

**Fig. 1.** Proposed mechanism for the reduction of enzyme-bound glycine to acetylphosphate

homododecamer containing a pyruvoyl residue on each subunit (m = 30 kDa). Coenzymes or a selenoprotein are not required (Hodgins and Abeles 1969; Seto and Stadtman 1976). This agrees well with the observation that growing cells of *C. sticklandii* are able to reduce proline but not glycine in the absence of selenite in the medium (Costilow 1977). This again shows that the selenoprotein is essential for this unique reductive-substrate chain phosphorylation. All other examples of this type of energy conservation are oxidative processes (Thauer et al. 1977).

## 2.2 *α,β-Elimination of Ammonia*

The α,β-elimination of a functional group from the α-position of a carboxylate requires the activation of the β-C-H bond by an additional keto, carboxyl, or thiolester group in order to remove the hydrogen as a proton. Among the amino acids occurring in proteins, only aspartate satisfies this condition since it is also a β-amino acid. Many organisms contain the enzyme aspartase which easily deaminates aspartate to fumarate without any cofactor or prosthetic group. The related deaminations of histidine and phenylalanine to urocanate and cinnamate, respectively, appear to be more complex, since the enzymes contain dehydroalanine as a prosthetic group which could facilitate the release of $NH_4^+$ (Givot et al. 1969; Hanson and Havir 1970). However, the proposed mechanisms do not explain the activation of the β-C-H bond by the dehydroalanine residue on the enzyme and by the aromatic residue on the substrate.

## 2.3 *Carbon-Carbon Rearrangement of Glutamate to 3-Methylaspartate*

Unlike that from aspartate, the α,β-elimination of the amino group from glutamate is not possible, since the β-C-H bond is not activated by a carboxyl or another electron withdrawing group. This problem is elegantly solved by *C. tetanomorphum* which readily ferments glutamate to acetate and butyrate via the methylaspartate pathway (Buckel and Barker 1974). In the first step (*S*)-glutamate is rearranged to (2*S*,3*S*)-3-methylaspartate from which the amino group is easily eliminated, whereby mesaconate (methylfumarate) is formed. During the study of these reactions, H.A. Barker discovered that an adenosyl derivative of pseudovitamin $B_{12}$ was required as coenzyme for this unusual rearrangement (Barker et al. 1958).

Notably, this was the first known function of this fascinating vitamin at the molecular level. Research with enzymes dependent on this coenzyme or on adenosylcobalamin (vitamin $B_{12}$ coenzyme), as well as with chemical model reactions indicated that a 5′-deoxyadenosine radical might be the active species in these rearrangements (Babior 1982; Dowd et al. 1988). This radical is generated by homolysis of the C-Co-bond in the coenzyme. The conversion of glutamate to 3-methylaspartate is initiated with the abstraction of the (4-*pro-S*) hydrogen of glutamate through the 5′-deoxyadenosine radical, whereby the methyl group of 5′-deoxyadenosine is formed as intermediate. The resulting glutamyl radical rearranges spontaneously to the corresponding 3-methylaspartyl radical. The final step is the reverse hydrogen transfer, whereby 3-methylaspartate is produced and the 5′-deoxyadenosine radical is regenerated (Fig. 2). This mechanism has the following remarkable features: (1) no exchange of the hydrogens with those of the solvent; (2) intermolecular migration of the hydrogen via 5′-deoxyadenosine; and (3) intramolecular migration of a glycyl residue in the opposite direction. An intermediate occurrence of a methylene radical on the branched amino acids was indicated by the use of (2*S*, 3*R* or 3*S*)-[3-$^2$H,3-$^3$H] glutamate as substrates. Thereby

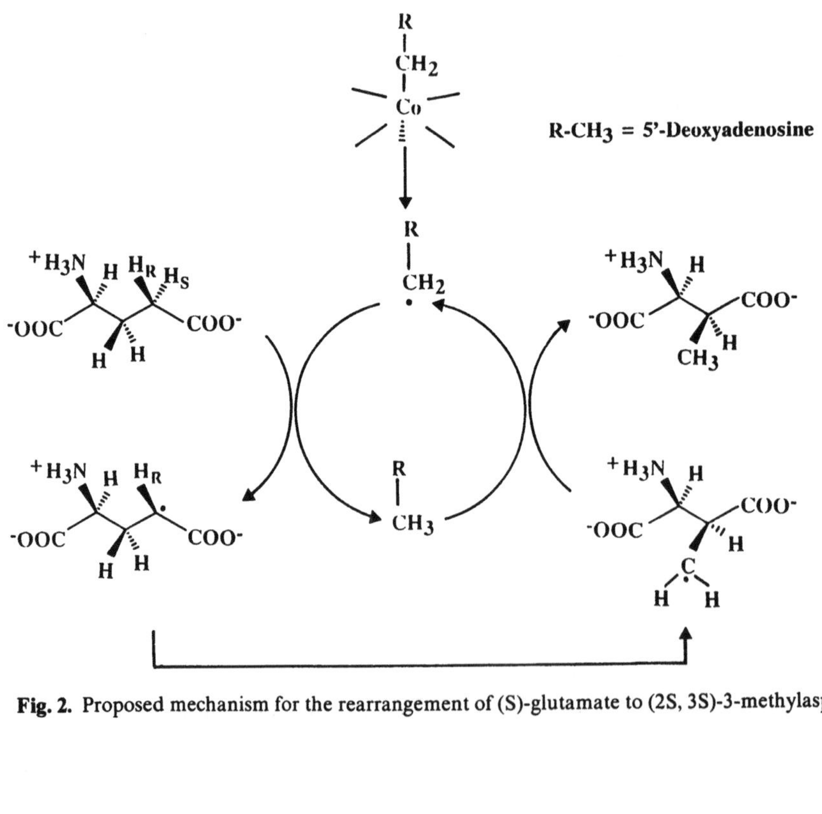


**Fig. 2.** Proposed mechanism for the rearrangement of (S)-glutamate to (2S, 3S)-3-methylaspartate

complete racemization was observed rather than the generation of optically active [$^1$H, $^2$H, $^3$H] methylgroups (Hartrampf and Buckel 1984).

### *2.4 Shift of the α-Amino Group to the β-Position*

Another possibility for the conversion of an α-amino acid into the β-isomer is a 2.3-shift of the amino group catalyzed by lysine (Fig. 3) and leucine amino mutases. The former enzyme is present in *C. subterminale* which readily ferments lysine to acetate and butyrate (Chirpich et al. 1970). The enzyme which has been obtained in an almost homogeneous form contains about 2 mol pyridoxal phosphate and 4 mol iron per mol of protein (m = 285 kDa). In addition 4 mol *S*-adenosylmethionine per mol of enzyme are required for the reaction. Although the enzyme is inactivated by oxygen, it is completely reactivated in the presence of $Fe^{2+}$ and a thiol under anaerobic conditions. During the reaction no exchange of the hydrogens at C-2 and C-3 with the solvent was observed. Further work on this unusual enzyme showed that the amino group migrates intramolecularly whereas the (3-*pro-R*) hydrogen moves intermolecularly from C-3 to C-2 (Aberhardt et al. 1983). Although these results are in complete agreement with adenosylcobalamin-dependent reactions, the lysine 2.3-aminomutase does not require this coenzyme. Therefore H.A. Barker suggested that *S*-adenosylmethionine might serve as a "poor man's adenosylcobalamine" (pers. commun. to Baker and Stadtman 1982). That this indeed could be right was recently shown by an experiment in which 10 nmol amino mutase transferred the complete radioactivity of 1 nmol *S*-[5′-$^3$H] adenosylmethionine to the equilibrium mixture of 70 μmol α- and β-lysine (Baraniak et al. 1989). Thus both 5′-methylene hydrogens of the 5′-position of the deoxyadenosyl moiety in *S*-adenosylmethionine participate in catalysis, indicating the intermediate formation of a methyl group, just like in adenosylcobalamin-dependent reactions. Furthermore, the iron in lysine 2.3-amino mutase could be a substitute for cobalt; finally, the migration of the amino group may be facilitated by Schiff base formation with the pyridoxal phosphate present in the enzyme. Interestingly, the further fate of β-lysine, the migration of the omega amino group from the 6- to the 5-position, yielding 3.5-diaminohexanoate, requires adenosylcobalamin (Baker and Stadtman 1982). Thus 5′-deoxyadenosine radicals may serve as the reactive agents in both amino group migrations although the sources of the radicals are very different. Contrary to lysine 2.3-amino mutase, the enzyme catalyzing the analogous conversion of leucine to β-leucine is adenosylcobalamin-dependent (Baker and Stadtman 1982).

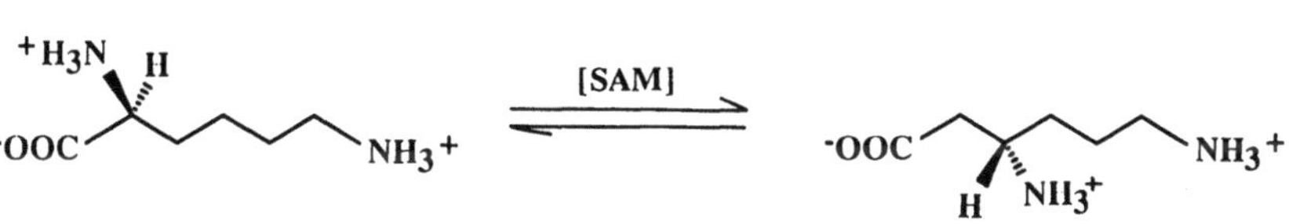


**Fig. 3.** *S*-Adenosylmethionine-dependent ([SAM]) migration of the α-aminogroup of lysine under the formation of β-lysine

*S*-Adenosylmethionine plays also an important role in the posttranslational modification of inactive pyruvate: formate lyase from *Escherichia coli*. Thereby the active enzyme containing a free radical is formed. The product of this conversion, 5′-deoxyadenosine, strongly indicates that the corresponding radical is an intermediate (Knappe et al. 1984).

### *2.5 Conversion to (R)-2-Hydroxyacyl-CoA and Dehydration to Enoyl-CoA*

The fermentation of α-amino acids via their corresponding (*R*)-2-hydroxy acids is the most common path to remove the α-amino group. In the first step, the amino acid is deaminated to the 2-oxo acid by oxidation, transamination, or dehydration. The following reduction by NADH affords the (*R*)-2-hydroxy acid which is activated to the CoA ester. Subsequent dehydration yields the enoyl-CoA which may be reduced to the saturated fatty acid. Well-studied examples for this pathway are the fermentations of alanine by *C. propionicum*, glutamate by *Acidaminococcus fermentans* (via the hydroxyglutarate pathway, see also the methylaspartate pathway, Sect. 3), phenylalanine, and leucine. There is also evidence that some bacteria ferment serine, cysteine, threonine, tyrosine, tryptophane, methionine, and histidine by this pathway (Schweiger et al. 1987).

The most interesting, but most elusive reaction of these fermentations is the reversible dehydration of (*R*)-2-hydroxyacyl-CoA to (*E*)-enoyl-CoA with a *syn*-geometry (Buckel 1980; Pitsch and Simon 1982). Thereby the C-H bond at C-3 which has to be cleaved is not activated by an adjacent electron withdrawing group such as a carboxylate or thiolester. It is the same problem as the elimination of ammonia from α-amino acids with the exception of aspartate (Sect. 2). Furthermore, in the reverse direction, water is added to the polarized double bond against the rule of Markovnikov. Therefore radicals were proposed as intermediates (Schweiger and Buckel 1984a; Kuchta and Abeles 1985).

Among these 2-hydroxyacyl-CoA dehydratases, only two have been purified catalyzing the dehydrations of (*R*)-lactyl-CoA to acrylyl-CoA in *C. propionicum* (Schweiger and Buckel 1984b, 1985; Kuchta and Abeles 1985; Kuchta et al. 1986) and of (*R*)-2-hydroxyglutaryl-CoA to glutaconyl-CoA in *A. fermentans* (Fig. 4; Buckel 1980; Schweiger and Buckel 1984a; Schweiger et al. 1987). Both are extremely oxygen-sensitive [Fe-S] proteins and have to be activated by ATP and $Mg^{2+}$ prior to the reaction. For this modification an additional, even more oxygen-sensitive protein is required. Inactivators are 0.1 mM 2.4-dinitrophenol, 0.2 mM chloramphenicol or 1 mM hydroxylamine (Buckel 1980; Schweiger and Buckel 1984b, 1985; Klees and Buckel unpubl.). The latter compound is effective

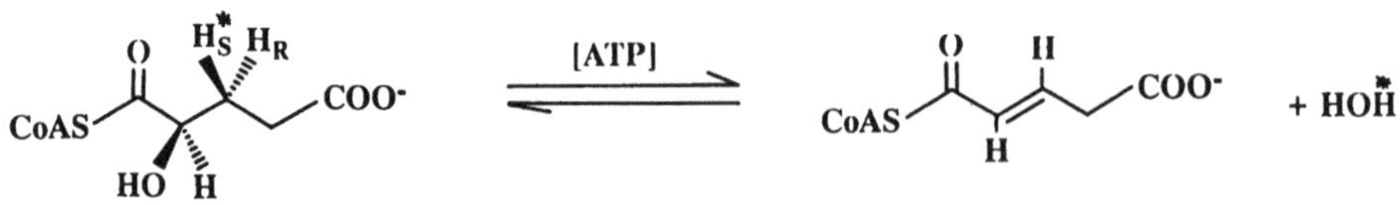


Fig. 4. ATP-dependent dehydration of (R)-2-hydroxyglutaryl-CoA to glutaconyl-CoA

at a concentration at which thiolesters remain uncleaved during the time of the assay. Perhaps these agents are radical scavengers.

The most intriguing problem of the mechanism appears to be the role of ATP in the activation of the dehydratases. With appropriately labeled ATP samples, no evidence for adenylation or phosphorylation of the enzyme was obtained. Another possibility could be a function similar to adenosylcobalamin (Sect. 3) or *S*-adenosylmethionine (Sect. 4), i.e., the generation of 5′-deoxadenosine radicals which could remove the hydrogen from the nonactivated C-H bond at C-3 of the substrates. Like lysine-2, 3-mutase, the 2-hydroxyacyl-CoA dehydratases contain iron which could be involved in the formation of this radical.

The 2-hydroxyglutaryl-CoA dehydratase from *A. fermentans* consists of two different subunits in a 1:1 ratio, $\alpha$, m = 53 870 Da and $\beta$, m = 41 857 Da. The native enzyme occurs in the cell-free extract as dimer ($\alpha\beta$), tetramer ($\alpha_2\beta_2$) or hexamer ($\alpha_3\beta_3$), whereas in the pure state only the tetramer is present. The enzyme contains 4 mol Fe and 4 mol inorganic sulfur per dimer (Schweiger et al. 1987). With the use of polyclonal antibodies and N-terminal sequences, the genes of both dehydratase subunits (hgdA for $\alpha$ and hgdB for $\beta$) were detected in lambda gt-11 and EMBL-3 gene banks. DNA-sequencing showed that gene A was followed very closely by gene B. The deduced primary structures of both genes showed no homology to other known proteins. Both subunits contain a high amount of cysteins (13 in $\alpha$ and 9 in $\beta$) which is unusual for prokaryotic proteins. The amino acids adjacent to the cysteines are often prolines or aromatic amino acids, which are also found in the MoFe subunit of the nitrogenases (Dutscho et al. 1989). However, these results gave no further insight into the mechanism of the enzyme. The main obstacle for studying this mechanism are the low activity (about 2 nkat $mg^{-1}$ protein) and the extreme instability of the purified 2-hydroxyglutaryl-CoA dehydratase, as well of its activator, even under strict anaerobic conditions.

## *2.6 Fermentation of Omega-Amino Acids*

The fermentations of the omega-amino acids 4-aminobutyrate and 5-aminovalerate are of interest, since the carboxylates are too far away in order to activate the C-H bond adjacent to the amino group necessary for $\alpha$, $\beta$-elimination of ammonia. 4-Aminobutyrate is fermented by *C. aminobutyricum* to ammonia, acetate and butyrate. In the first two steps the amino acid is converted to 4-hydroxybutyrate, which is supposed to be activated to the CoA ester and dehydrated to vinylacetyl-CoA (Hardman and Stadtman 1963). Thus the same mechanistic problem arises as with the dehydration of 2-hydroxyglutaryl-CoA, as discussed above. However, recent data show that one of the nonactivated hydrogens at C-3 of 4-hydroxy[3-$^3$H]butyryl-CoA is indeed removed stereospecifically, which agrees well with the hypothetical pathway (Fig. 5). The extreme oxygen sensitivity of the dehydratase may indicate the involvement of a transition metal and perhaps a radical in the mechanism (Willadsen and Buckel 1990).

*C. aminovalericum* ferments 5-aminovalerate in a similar manner via glutarsemialdehyde and 5-hydroxyvalerate to ammonia, acetate, propionate, and valerate. Again, 5-hydroxyvalerate is activated to the CoA ester and dehydrated to

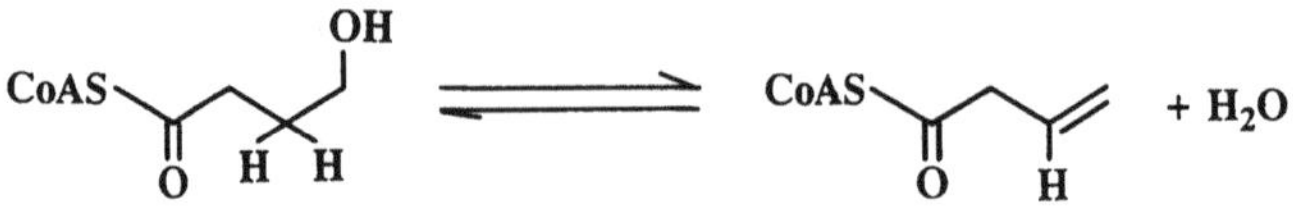


**Fig. 5.** Dehydration of 4-hydroxybutyryl-CoA to vinylacetyl-CoA

4-pentenoyl-CoA (Barker et al. 1987). However, the green dehydratase, which was recently purified to homogeneity, contains FAD and is not sensitive towards oxygen. The additional acyl-CoA dehydrogenase activity of the enzyme indicates a very different mechanism in which an $\alpha,\beta$-double bond is transiently formed. The intermediate (*E*)-5-hydroxy-2-pentenoyl-CoA, a vinylogue of a 3-hydroxyacyl-CoA, is readily dehydrated to 2.4-pentadienoyl-CoA, which is reduced to the final product 4-pentenoyl-CoA (Fig. 6). Furthermore, if the hydration of 4-pentenoyl-CoA is performed in $D_2O$, one deuterium atom is incorporated at C-2 in addition to one at C-4 of 5-hydroxyvaleryl-CoA. All three reactions, oxidation, dehydration and reduction are catalyzed by this single enzyme (Eikmanns and Buckel unpubl.).

## 3 Summary and Outlook

This chapter shows chemical problems in the fermentation of amino acids which anaerobic bacteria have to cope with. There are several reactions of great biochemical interest whose mechanisms are hardly understood: the reduction of glycine to acetyl phosphate; the rearrangement of glutamate to 3-methylaspartate; the shift of the $\alpha$-amino group of lysine to the $\beta$-position; and the dehydrations of 2- as well as of 4-hydroxyacyl-CoA derivatives. One important agent of these reactions appears to be the 5′-deoxyadenosine radical, for which formation three methods have probably been developed by evolution. The possible precursors of this radical increase in complexity from ATP via *S*-adenosylmethionine to

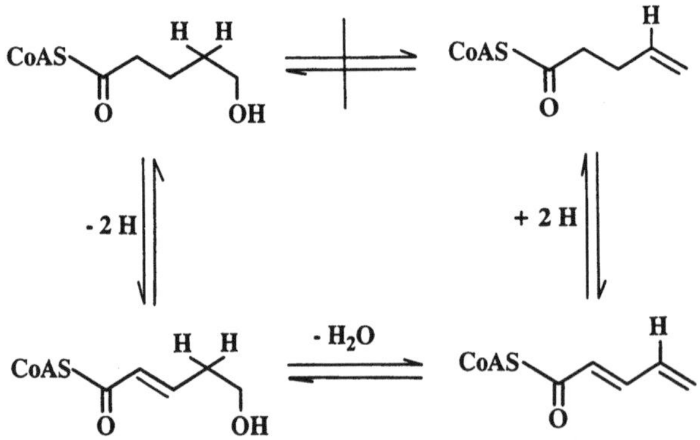


**Fig. 6.** Mechanism of the dehydration of 5-hydroxyvaleryl-CoA to 4-pentenoyl-CoA

adenosylcobalamin. The stability of the involved enzymes towards oxygen increases in the same order. Perhaps evolution has taken a similar path.

In the fermentation of amino acids, the complete set of hydroxyacyl-CoA dehydrations from the 2- to the 5-derivative occurs. The dehydrations of 3- and 5-hydroxyacyl-CoA have ionic mechanisms, whereas in those of 2- and 4-hydroxyacyl-CoA, radicals are most probably involved. With the exception of $\beta$-amino acids, all other amino acids are converted to the corresponding hydroxyacids prior to the elimination. There is no obvious chemical reason for this.

*Acknowledgments.* The author's work described in this paper was supported by funds from the Deutsche Forschungsgemeinschat and the Fonds der Chemischen Industrie. The author thanks Professor R. K. Thauer (Philipps Universitat Marburg) for helpful discussions.

## References

Aberhart, D.J., Gould, S.J., Lin, H.-J., Thiruvengadam, T.K. & Weiller, B.H. (1983) Stereochemistry of lysine 2,3-aminomutase isolated from *Clostridium subterminale* strain SB4. J. Am. Chem. Soc. 105:5461–5470

Arkowitz, R.A. & Abeles, R.H. (1989) Identification of acetyl phosphate as the product of clostridial glycine reductase: evidence for an acyl enzyme intermediate. Biochemistry 28:4639–4644

Arkowitz, R.A. & Abeles, R.H. (1990) Isolation and characterisation of a covalent selenocysteine intermediate in the glycine reductase system. J. Am. Chem. Soc. 112:870–872

Babior, B.M. (1982) Ethanolamine ammonia-lyase. In: Dolphin, D. (ed) $B_{12}$, Vol. 2. Wiley, New York, pp. 263–287

Baker, J.J. & Stadtman, T.C. (1982) Amino mutases. In: Dolphin, D. (ed) $B_{12}$, Vol. 2. Wiley, New York, pp. 203–232

Baraniak, J., Moss, M.L. & Frey, P.A. (1989) Lysine 2,3-aminomutase. J. Biol. Chem. 264:1357–1360

Barker, H.A. (1937) On the fermentation of glutamic acid. Enzymologia 2:175–182

Barker, H.A. (1961) Fermentations of nitrogenous organic compounds. In: Gunsalus, I.C., Stanier, R.Y. (eds.) The bacteria. Academic Press, Lond New York, pp. 151–207

Barker, H.A. (1981) Amino acid degradation by anaerobic bacteria. Ann. Rev. Biochem. 50:23–40

Barker, H.A., Weißbach, H. & Smyth, R.D. (1958) A coenzyme containing pseudovitamin $B_{12}$. Proc. Natl. Acad. Sci. USA 44:1093

Barker, H.A., D'Ari, L. & Kahn, J. (1987) Enzymatic reactions in the degradation of 5-aminovalerate by *Clostridium aminovalericum*. J. Biol. Chem. 262:8994–9003

Barnard, G.F. & Akhtar, M. (1979) Mechanistic and stereochemical studies on the glycine reductase of *Clostridium sticklandii*. Eur. J. Biochem. 99:593–603

Buckel, W. (1980) The reversible dehydration of (*R*)-2-hydroxyglutarate to (*E*)-glutaconate. Eur. J. Biochem. 106:439–447

Buckel, W. & Barker, H.A. (1974) Two pathways of glutamate fermentation by anaerobic bacteria. J. Bacteriol. 117:1248–1260

Cardon, B.P. & Barker, H.A. (1946) Two new amino-acid-fermenting bacteria, *Clostridium propionicum* and *Diplococcus glycinophilus*. J. Bacteriol. 52:629–634

Chirpich, T.P., Zappia, V., Costilow, R.N. & Barker, H.A. (1970) Lysine 2,3-aminomutase. Purification and properties of a pyridoxal phosphate and *S*-adenosylmethionine-activated enzyme. J. Biol. Chem. 245:1778–1789

Cone, J.E., Martin del Rio, R., Davis, J.N. & Stadtman, T.C. (1976) Chemical characterisation of the selenoprotein component of clostridial glycine reductase: identification of selenocysteine as the organoselenium moiety. Proc. Natl. Acad. Sci. USA 73:2659–2663

Costilow, R.N. (1977) Selenium requirement for the growth of *Clostridium sporogenes* with glycine as the oxidant in Stickland reaction systems. J. Bacteriol. 131:366–368

Dowd, P., Choi, S.-C., Duah, F. & Kaufman, C. (1988) A model for the enzyme-catalyzed, coenzyme $B_{12}$-dependent interconversion of $\beta$-methylaspartate with glutamate. Tetrahedron 44:2137–2148

Dutscho, R., Wohlfarth, G., Buckel, P. & Buckel, W. (1989) Cloning and sequencing of the genes of 2-hydroxyglutaryl-CoA dehydratase from *Acidaminococcus fermentans*. Eur. J. Biochem. 181:741–746

Givot, I.L., Smith, T.A. & Abeles, R.H. (1969) Studies on the mechanism of action and the structure of the electrophilic center of histidine ammonia lyase. J. Biol. Chem. 244:6341–6353

Hanson, K.R. & Havir, E.A. (1970) L-Phenylalanine ammonia-lyase. IV. Evidence that the prosthetic group contains a dehydroalanyl residue and mechanism of action. Arch. Biochem. Biophys. 141:1–17

Hardman, J.K. & Stadtman, T.C. (1963) Metabolism of omega-amino acids. IV. Gamma-Aminobutyrate fermentation by cell-free extracts of *Clostridium aminobutyricum*. J. Biol. Chem. 238:2088–2093

Hartrampf, G. & Buckel, W. (1984) The stereochemistry of the formation of the methyl group in the glutamate mutase-catalysed reaction in *Clostridium tetanomorphum*. FEBS Lett. 171:73–78

Hodgins, D.S. & Abeles, R.H. (1969) Studies of the mechanism of action of D-proline reductase: the presence on covalently bound pyruvate and its role in the catalytic process. Arch. Biochem. Biophys. 130:274–285

Knappe, J., Neugebauer, F.A., Blaschkowski, H.P. & Gänzler, M. (1984) Post-translational activation introduces a free radical into pyruvate formate-lyase. Proc. Natl. Acad. Sci. USA 81:1332–1335

Kuchta, R.D. & Abeles, R.H. (1985) Lactate reduction in *Clostridium propionicum*. J. Biol. Chem. 260:13181–13189

Kuchta, R.D. Hanson, G.R., Holmquist, B. & Abeles, R.H. (1986) Fe-S centers in lactyl-CoA dehydratase. Biochemistry 25:7401–7307

Pitsch, C. & Simon, H. (1982) The stereochemical course of the water elimination from (2*R*)-phenyllactate in the amino acid fermentation of *Clostridium* sporogenes. Hoppe Seyler's Z. Physiol. Chem. 363:1253–1257

Schweiger, G. & Buckel, W. (1984a) Studies on the dehydration of (*R*)-2-hydroxyglutarate in *Acidaminococcus fermentans*. A radical mechanism? Arch. Microbiol. 137:302–307

Schweiger, G. & Buckel, W. (1984b) On the dehydration of (*R*)-lactate in the fermentation of alanine to propionate by *Clostridium propionicum*. FEBS Lett. 171:79–84

Schweiger, G. & Buckel, W. (1985) Identification of acrylate, the product of the dehydration of (*R*)-lactate catalysed by cell-free extracts from *Clostridium propionicum*. FEBS Lett. 185:253–256

Schweiger, G., Dutscho, R. & Buckel, W. (1987) Purification of 2-hydroxyglutaryl-CoA dehydratase from *Acidaminococcus fermentans*. An iron-sulfur protein. Eur. J. Biochem. 169:441–448

Seto, B. & Stadtman, T.C. (1976) Purification and properties of proline reductase from *Clostridium sticklandii*. J. Biol. Chem. 251:2453–2439

Stickland, H.L. (1934) Studies in the metabolism of the strict anaerobes (genus *Clostridium*). I. The chemical reactions by which *Cl. sporogenes* obtains its energy. Biochem. J. 2:1746–1759

Tanaka, H. & Stadtman, T.C. (1979) Selenium-dependent clostridial glycine reductase. J. Biol. Chem. 254:447–452

Thauer, R.K. Jungermann, K. & Decker, K. (1977) Energy conservation in chemotrophic anaerobic bacteria. Bacteriol. Rev. 41:100–180

Whiteley, H.R. (1957) Fermentation of amino acids by *Micrococcus aerogenes*. J. Bacteriol. 74:324–330

Willadsen, P. & Buckel, W. (1990) Assay of 4-hydroxybutyryl-CoA dehydratase from *Clostridium aminobutyricum*. FEMS Microbiol. Lett. 70:187–192

Zindel, U., Freudenberg, W., Rieth, M., Andreesen, J.R., Schnell, J. & Widdel, F. (1988) *Eubacterium acidaminophilum*. sp. nov., a versatile amino acid-degrading anaerobe producing or utilizing $H_2$ or formate. Arch. Microbiol. 150:254–266

# Anaerobic Pyruvate Metabolism in Enterobacteria: The Radical Enzyme, Pyruvate Formate-Lyase

J. KNAPPE[1]

## 1 Introduction

*Escherichia coli* as well as many other facultative anaerobic organisms, traditionally classified as enterobacteria, are capable of degrading glucose into ethanol, acetic acid and formic acid as an energy source for cell growth in the absence of oxygen. This characteristic pattern of fermentation products results from a special processing of the pyruvate intermediate formed from the Embden-Meyerhof route: thiolysis of its carbon chain by CoA, yielding acetyl-CoA and formate. *E. coli* cells can even be grown anaerobically on pyruvate as sole substrate, in which case the ATP energy stems from substrate-level phosphorylation along the three-step mini-fermentation route, pyruvate → acetate + formate (see Fig. 1).

The crucial pyruvate cleavage step:

$$\text{pyruvate} + \text{CoA} \rightleftharpoons \text{acetyl-CoA} + \text{formate},$$

discovered almost half a century ago in the laboratories of Werkman and Lipmann (Kalnitsky and Werkman 1943; Utter et al. 1944), is unique in the catalogue of chemical reactions of living organisms. Its mechanism became apparent only in 1984 with the discovery that the catalyst, pyruvate formate-lyase (PFL), has the fascinating feature of containing a stable organic-free radical intimately integrated in the protein structure (Knappe et al. 1984). In this chapter, I will briefly discuss structure/function aspects of the enzyme from *E. coli* studied in our laboratory. A more detailed account of this system can be found in a recent review article (Knappe and Sawers 1990).

## 2 Catalytic Function of the Protein Radical

Our studies of the mechanism of pyruvate formate-lyase first revealed an indirect group transfer reaction mode wherein the SH group of a cysteinyl residue (Cys-419) operates covalent-catalytically (Knappe et al. 1974). The overall process is divided into two distinct half-reactions which are linked through an isolatable acetyl-enzyme intermediate:

$$\text{pyruvate} + \text{HS-E} \rightleftharpoons \text{acetyl-S-E} + \text{formate}$$
$$\text{acetyl-S-E} + \text{CoA} \rightleftharpoons \text{HS-E} + \text{acetyl-CoA}.$$

[1]Institut für Biologische Chemie, Im Neuenheimer Feld 501, W 6900 Heidelberg, FRG

41. Colloquium Mosbach 1990
The Molecular Basis of Bacterial Metabolism

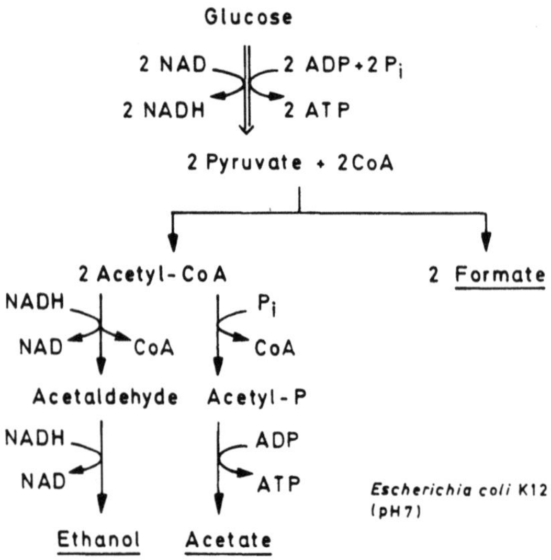


**Fig. 1.** Fermentation of glucose in *Escherichia coli*

The individual reactions are readily reversible. Remarkably high turnover numbers for both the cleavage as well as synthesis directions reflect the outstanding effectiveness of pyruvate formate-lyase (Table 1).

This general reaction mode is formally quite similar to that of $\beta$-keto thiolase of the fatty acid oxidation cycle. With respect to the chemical mechanism, however, it had always been extremely difficult to believe that the critical first half-reaction of the carbon-carbon bond cleavage/synthesis could occur by a common Claisen-type reaction involving carbanion chemistry as it applies to the thiolase process. That, in fact, a homolytic rather than heterolytic mechanism would apply for the pyruvate formate-lyase process became then highly suggestive with the discovery of the radical content of the enzyme. Figure 2 depicts a feasible minimum mechanism of the first half-reaction, wherein the protein radical functions as a

**Table 1.** Characteristics of pyruvate formate-lyase from *E. coli*

| | | | |
|---|---|---|---|
| Structure | (Conradt et al. 1984; Rödel et al. 1988; Unkrig et al. 1989): Homodimer from a 759 aa-polypeptide. Active form contains 1 spin (EPR signal at g = 2.0037) | | |
| Catalytic parameters (30 °C, pH 8) (Knappe et al. 1974): | | | |
| | $k_{cat}$ ($s^{-1}$) | $K_m$ | (mM) |
| Forward reaction | 770 | Pyruvate<br>CoA | 2<br>0.007 |
| Reverse reaction | 260 | Formate<br>Acetyl-CoA | 24<br>0.05 |

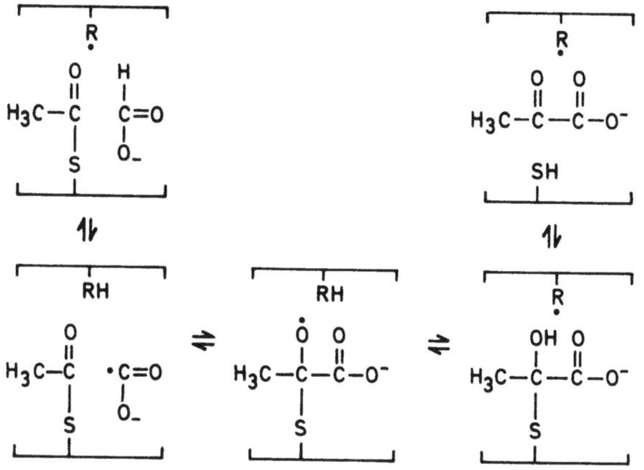

**Fig. 2.** Proposed mechanism of pyruvate formate-lyase (first half-reaction of the catalytic cycle). *SH* The covalent-catalytic Cys-419 residue; *Ṛ*, *RH* the putative amino acid radical and its quenched form, respectively (Plaga et al. 1988)

hydrogen-atom transferring coenzymatic element. C-C bond synthesis is proposed to occur by the addition of the formate radical species ($\dot{C}O_2^-$) to the carbonyl group of the acetyl moiety. This should be reasonable, since carbon-centered radicals are well-established nucleophiles in organic chemistry.

Clearly, it is questionable whether any substrate-based radical intermediates in the PFL reaction could ever be observed in a direct manner. But we have recently obtained experimental proof of the functional role of the protein radical through studies on hypophosphite, which turned out to be a suicide substrate of pyruvate formate-lyase (Plaga et al. 1988). As a close analogue of formate, hypophosphite reacts covalently with the acetyl-enzyme intermediate, producing a carbon-phosphorus bond (Fig. 3). The dead-end product is 1-hydroxyethylphosphonate which is linked as thioester to a distinct cysteinyl residue (Cys-418) that is adjacent to the acetyl-carrying Cys-419. This structure was determined by $^{31}$P-NMR

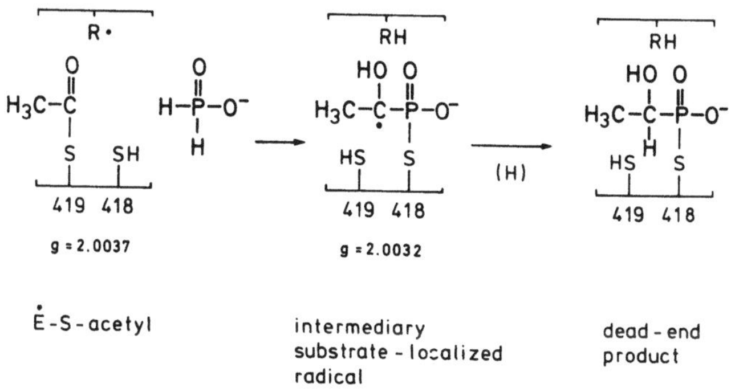


**Fig. 3.** Mechanism-based inactivation of pyruvate formate-lyase by hypophosphite (Plaga et al. 1988; Unkrig et al. 1989)

spectroscopy and by phosphodiesterase digestion of the modified protein, which released the free hydroxyethyl-phosphonate compound. Interestingly, when we applied EPR spectroscopy to the analogue reaction with the acetyl enzyme, now carried out at the lowered temperature of 0° C, we found that the EPR signal of the protein radical (see Fig. 4A) was rapidly replaced by a new EPR spectrum (Unkrig et al. 1989). This was unambiguously assigned, via computer simulation of the spectrum and by isotopic substitutions, to the α-phosphoryl radical precursor compound of the dead-end product (see Fig. 3). With a half-life of 20 min (at 0° C), this intermediate is remarkably stable, which must be due to it being buried in the native protein fold of the active site. Whatever the mechanistic route of the analogue reaction might be, the occurrence of a pseudosubstrate-localized radical is conclusive evidence of the radical-chemical principle of PFL catalysis.

An intriguing question posed by the results of the analogue reaction is whether the further cysteinyl residue Cys-418 is of covalent-catalytic relevance for the normal reaction cycle of PFL. This is being currently investigated by mutagenesis experiments.

## 3 Structural Characteristics of the Protein Radical

The chemical structure of the protein radical is still unsettled. The *g*-value of the EPR signal and the microwave power saturation behaviour are typical of an organic-free radical and we have some albeit circumstantial evidence that the unpaired electron is localized on a common amino acid residue of the polypeptide chain (Unkrig et al. 1989).

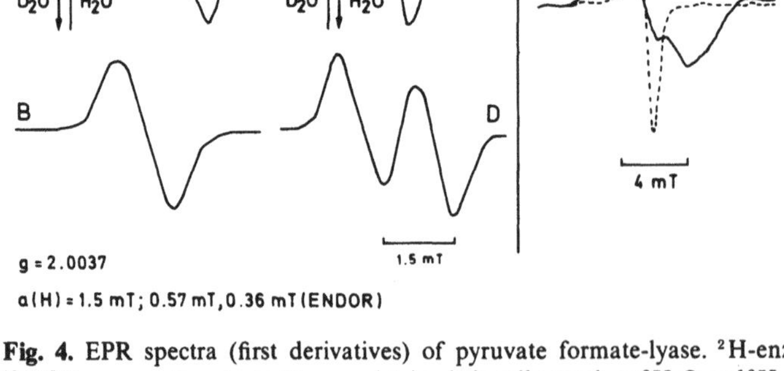


**Fig. 4.** EPR spectra (first derivatives) of pyruvate formate-lyase. $^2$H-enzyme (spectrum *C*) and $^{13}$C/$^2$H-enzyme (spectrum *E*) were obtained via cell growth on $^2H_2O$ and $^2H_2O$/$^{13}C_6$-glucose medium, respectively. Spectra *B* and *D* were registered after gel filtration of the enzyme samples into $D_2O$ or $H_2O$ buffer. The *dotted line* in *E* indicates the EPR spectrum of $^{12}$C/$^2$H-enzyme (*C*) (Unkrig et al. 1989)

To determine its identity we have carried out various isotopic substitutions of the enzyme protein. By first replacing all hydrogens by deuterium, via cell growth on $D_2O$, it became apparent that the asymmetric doublet structure of the EPR spectrum of the normal protein occurs by couplings of the unpaired electron to a variety of hydrogen nuclei. As depicted in Fig. 4, the particular hydrogen that gives rise to the major doublet splitting (1.5 mT) turned out to be a solvent-exchangeable proton. Other isotopic substitutions, which employed $^{15}N$ (I = 1/2) and $^{33}S$ (I = 3/2), made clear that the unpaired electron is neither nitrogen- nor sulfur-centered. Unfortunately, incorporation experiments, undertaken with a series of deuterated amino acids, have not given any clue so far as to the identity of the radical moiety. Therefore, $^{13}C$ (I = 1/2) is now being used as isotopic probe. The distinctive, very broad EPR signal of the [$^{13}C$]enzyme (Fig. 4, spectrum E) indicates couplings of the unpaired electron to a central carbon nucleus as well as adjacent carbon nuclei in $\alpha$- or $\beta$-positions. Hence, by systematic incorporation of selectively carbon-labelled amino acids (with the use of the corresponding auxotrophic mutants), the amino acid radical should not escape its delineation. Presently, our data point to an aliphatic rather than aromatic amino acid.

Spin-trapping is another strategy for characterizing radicals, but spin-trap reagents commonly used in organic chemistry (e.g., DMPO) do not react with the protein radical of pyruvate formate-lyase. Clearly, the radical site is inaccessible to unspecific, bulky molecules. In contrast, oxygen ($O_2$) has ready access and rapidly destroys the radical and the catalytic activity (Knappe et al. 1984). Most intriguingly, the 85 kDa polypeptide chain is concomitantly cleaved into 82 and 3 kDa fragments. This probably occurs through a peroxyl-radical intermediate, i.e., according to mechanisms discussed in $\gamma$-radiolysis research, where oxygen-mediated protein fragmentation is a well-known phenomenon (see Wolff et al. 1986). We have recently found (V. Wagner, M. Frey and J. Knappe, unpubl. results) that the large fragment represents the N-terminal portion of the original polypeptide, terminating with Val-732 as the last intact C-terminal amino acid residue. The 3 kDa fragment is not amenable to Edman degradation. Although the chemical modification occurring with the oxygen cleavage remains to be elucidated, our findings strongly indicate that the radical-bearing amino acid is located closely downstream of Val-732; that is, Ser-733 would be a likely candidate (Fig. 5).

Besides determining the covalent structure, the topology of the protein radical should be a further intriguing aspect. The following questions are relevant: (1) Is the radical in immediate proximity to the covalent-catalytic Cys-419; or is it perhaps quite remote, with an interspersed H-atom transfer relais system, which

acetyl transfer site — putative radical site

1 — 419 — 733 — 759

$H_2N$-Ser------Cys-Cys-Val-----Val-Ser-Gly---Met-COOH

**Fig. 5.** Primary-structural distribution of functional sites of pyruvate formate-lyase

would be triggered upon substrate binding? (2) Which elements of the native protein fold achieve the stabilization of the (resting state) radical, thus preventing spontaneous quenching? It is important to note that pyruvate formate-lyase does not contain a metal as this occurs with the iron-containing tyrosyl radical center of *E. coli* ribonucleotide reductase (Reichard and Ehrenberg 1983). (In this prototype of radical enzymes the metal is believed to contribute to the thermal stabilization of the organic free radical.) Clearly, the X-ray structure analysis of pyruvate formate-lyase, which has recently been started, is expected to provide deeper insight into such questions.

## 4 Post-Translational Interconversion of Pyruvate Formate-Lyase and Transcriptional Control of the *PFL* Gene

The synthetic system that produces the unpaired electron state of pyruvate formate-lyase in *E. coli* cells was disclosed prior to the identification of this novel modification. Figure 6 summarizes the various enzymes and cofactors which were found in our laboratory to be directly or indirectly involved (see Knappe and Sawers 1990).

Radical introduction is mediated by a monomeric 28 kDa protein, termed "activase", which contains one bound ferrous ion and employs reduced flavodoxin and S-adenosyl methionine as cosubstrates. With each protein radical produced, adenosyl methionine is converted stoichiometrically to 5′-deoxyadenosine and methionine (Knappe et al. 1984; Unkrig et al. 1989). A close analogy of this system to adenosyl cobalamin-involving reactions (see Stubbe 1989) is quite obvious and suggests an enzyme-bound Fe-adenosyl intermediate serving as the H-atom abstracting element. We have no evidence so far for such a mechanism. But the

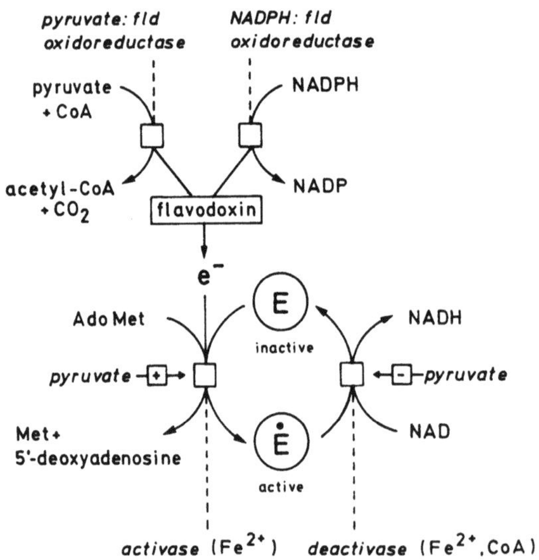


**Fig. 6.** Post-translational system of pyruvate formate-lyase interconversion between radical ($\dot{E}$) and non-radical ($E$) forms. [+] and [−] indicate positive and negative effects of pyruvate on the conversion reactions

extra hydrogen of 5′-deoxyadenosine could indeed stem from the protein substrate, since running the reaction in $D_2O$ buffer medium yields unlabelled 5′-deoxyadenosine.

It is interesting to note that the activase reaction is a unique mode of the *net* synthesis of a stable protein radical, in contrast to the transient substrate radicals occurring in $B_{12}$ systems.

More recently, the presence of a distinct "deactivase" in *E. coli* cells, also an Fe(II) protein, has been established (Kessler and Knappe 1988). It catalyzes the quenching of the radical of pyruvate formate-lyase in the presence of CoA and NAD. NADH formation is involved in this process, the mechanism and redox chemistry of which remains to be elucidated. Structure analyses, including electron microscopy, revealed that deactivase is a polymeric protein of intriguing architecture. About 40 protomers of 100 kDa are helically assembled into rods of some 130 nm in length (D. Kessler, W. Herth and J. Knappe, unpubl. results). Our data strongly indicate that deactivase is identical with the "spirosome", a proteinaceous ultrastructural element previously detected in various obligate and facultative anaerobes (Kawata et al. 1976; Matayoshi et al. 1989).

Estimated from the in vitro catalytic data of the two converter enzymes, the total pool of pyruvate formate-lyase present in the anaerobic cell could be converted to the radical form as well as back to the inactive non-radical form, each in a time period of about 1 min. But direct measurements have shown that the enzyme is virtually completely in the active form during anaerobic cell growth (at pH 7). Conversely, it is kept totally in the non-radical form during aerobiosis, which is reasonable in view of the destructive effect of oxygen on the protein radical. Clearly, the interconversion cycle must be subject to tight control, which is likely to involve further metabolic factors beyond the diametrical pyruvate effects found so far (see Fig. 6).

Another intriguing aspect of the PFL system of *E. coli* concerns the regulation of the gene transcription. As summarized in Fig. 7, pyruvate formate-lyase as well as the deactivating enzyme are anaerobically induced proteins, whereas activase is synthesized constitutively. Sawers and Böck (1989) have recently delineated as many as seven promoters, covering a region of 1.2 kb of DNA, which are necessary for expression of the *pfl* gene. They are coordinately regulated by anaerobiosis, and this is to a large extent due to the regulator protein Fnr. The 5′-flanking region of *pfl* encodes a further protein (orf), which is synthesized only to a very small level. Its function is unknown; it is definitely not essential for generating active pyruvate formate-lyase.

The present distribution of the PFL system, although not yet systematically screened, goes far beyond the group of enterobacteria; it is even found in unicellular green algae (for references, see Knappe and Sawers 1990). The unique mechanism of this enzyme suggests that it has evolved along lines that were independent of the occurrence of the other two enzyme systems known to transform pyruvate to acetyl-CoA; i.e. the pyruvate dehydrogenase complex and the pyruvate:flavodoxin (ferredoxin) oxidoreductase. From a certain chemical point of view, the radical of pyruvate formate-lyase represents a coenzymatic alternative to thiamine diphosphate for processing $\alpha$-keto carboxylic acids.

**A**

| | Copy numbers of polypeptide chains | |
|---|---|---|
| | aerobic | anaerobic |
| PYRUVATE FORMATE - LYASE | 3000 | 35000 |
| ACTIVATING ENZYME | 2000 | 2000 |
| DEACTIVATING ENZYME | 3000 | 30000 |

**B**

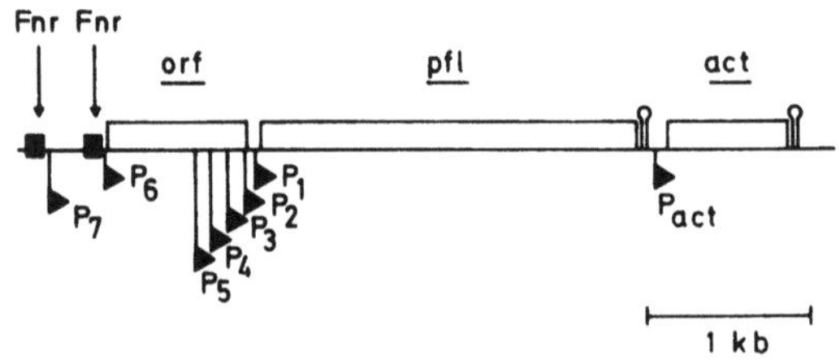


**Fig. 7.** **A** Enzyme levels and **B** transcriptional organization of the *pfl* gene. *Figures* in **A** denote estimated copy numbers of polypeptide chains in the K12 wild-type cell under aerobic and anaerobic growth conditions (Knappe and Sawers 1990). $P_{1\text{-}7}$ and $P_{act}$ indicate the positions of *pfl* and *act* (activase) gene promoters: signify transcription terminators: *filled boxes* denote Fnr protein recognition sequences (Rödel et al. 1988; Sawers and Böck 1989; Sauter and Sawers 1990)

## References

Conradt, H., Hohmann-Berger, M., Hohmann, H.P., Blaschkowski, H.P. & Knappe, J. (1984) Pyruvate formate-lyase (inactive form) and pyruvate-formate-lyase-activating enzyme of *Escherichia coli*: isolation and structural properties. Arch. Biochem. Biophys. 228:133–142

Kalnitsky, G. & Werkman, C.H. (1943) The anaerobic dissimilation of pyruvate by a cell-free extract of *Escherichia coli*. Arch. Biochem. Biophys. 2:113–124

Kawata, T., Masuda, K. & Ueki, Y. (1976) Isolation and ultrastructural characterization of fine spirals (spirosome) from *Lactobacillus brevis*. J. Electron Microsc. 25:283–288

Kessler, D. & Knappe, J. (1988) Enzymatic deactivation of the radical enzyme pyruvate formate-lyase. 14th Int. Congr. Biochem. (Prague) Abstract TH 367

Knappe, J. & Sawers, G. (1990) A radical-chemical route to acetyl-CoA: the anaerobically induced pyruvate formate-lyase system of *Escherichia coli*. FEMS Microbiol. Rev. (in press)

Knappe, J., Blaschkowski, H.P., Gröbner, P. & Schmitt, T. (1974) Pyruvate formate-lyase of *Escherichia coli*: the acetyl-enzyme intermediate. Eur. J. Biochem. 50:253–263

Knappe, J., Neugebauer, F.A., Blaschkowski, H.P. & Gänzler, M. (1984) Post-translational activation introduces a free radical into pyruvate formate-lyase. Proc. Natl. Acad. Sci. USA 81:1332–1335

Matayoshi, S., Oda, H. & Sarwar, G. (1989) Relationship between the production of spirosomes and anaerobic glycolysis in *Escherichia coli*. B. J. Gen. Microbiol. 135:525–529

Plaga, W., Frank, R. & Knappe, J. (1988) Catalytic-site mapping of pyruvate formate-lyase. Hypophosphite reaction on the acetyl-enzyme intermediate affords carbon-phosphorus bond synthesis (1-hydroxyethylphosphonate). Eur. J. Biochem. 178:445–450

Reichard, P. & Ehrenberg, A. (1983) Ribonucleotide reductase – a radical enzyme. Science 221:514–519

Rödel, W., Plaga, W., Frank, R. & Knappe, J. (1988) Primary structures of *Escherichia coli* pyruvate formate-lyase and pyruvate-formate-lyase-activating enzyme deduced from the DNA nucleotide sequences. Eur. J. Biochem. 177:153–158

Sauter, M. & Sawers, R.G. (1990) Transcriptional analysis of the gene encoding pyruvate formate-lyase-activating enzyme of *Escherichia coli*. Mol. Microbiol. 4:355–364

Sawers, G. & Böck, A. (1989) Novel transcriptional control of the pyruvate formate-lyase gene: upstream regulatory sequences and multiple promoters regulate anaerobic expression. J. Bacteriol. 171:2485–2498

Stubbe, J.A. (1989) Protein radical involvement in biological catalysis? Annu. Rev. Biochem. 58:257–285

Unkrig, V., Neugebauer, F.A. & Knappe, J. (1989) The free radical of pyruvate formate-lyase. Characterisation by EPR spectroscopy and involvement in catalysis as studied with the substrate-analogue hypophosphite. Eur. J. Biochem. 184:723–728

Utter, M.F., Werkman, C.H. & Lipmann, F. (1944) Reversibility of the phosphoroclastic split of pyruvate. J. Biol. Chem. 154:723–724

Wolff, S.P., Gamer, A. & Dean, R.T. (1986) Free radicals, lipids and protein degradation. Trends Biochem. Sci. 11:27–31

# Hydrogen Metabolism: Structure and Function of Hydrogenases

S.P.J. ALBRACHT[1]

## 1 Introduction

Hydrogenases are involved in the reduction of $H^+$ or the oxidation of $H_2$ in a wide variety of microorganisms. Anaerobic bacteria living on the fermentation of organic substrates often dispose of their excess of reducing equivalents by way of the reduction of protons. The major source of these reducing equivalents is pyruvate. Many bacteria can oxidize $H_2$, which enables them to produce ATP and to reduce a variety of substrates. Hydrogen is also produced in $N_2$-fixing organisms: at least one $H_2$ is formed for every molecule of $N_2$ reduced. In such organisms hydrogenase functions to recapture the energy-rich reducing equivalents. For more extensive information on the function of hydrogenases, the reader is referred to the impressive general review of Adams et al. (1980). The discovery of three genetically different hydrogenases in *Desulfovibrio vulgaris* (Lissolo et al. 1986) has reinforced the notion (Odom and Peck 1981) that molecular hydrogen also plays a central role in the bioenergetics of sulphate-reducing bacteria (Fauque et al. 1988).

Two classes of hydrogenases can now be recognized: (1) Fe hydrogenases, also called "Fe-only" hydrogenases, containing $[4Fe\text{-}4S]^{2+(2+;1+)}$ clusters and a novel 6Fe cluster; and (2) Ni hydrogenases. Either S, or S plus Se, occurs as ligands to nickel. The Ni-S and Ni-S-Se enzymes in *D. vulgaris* are encoded by distinct genes (Li et al. 1987) and are considered as subclasses. Iron-sulphur clusters of the $[4Fe\text{-}4S]^{2+(2+;1+)}$ type and the $[3Fe\text{-}4S]^{1+(1+;0)}$ type can be found here. Iron hydrogenases are 10–100 times as active as Ni hydrogenases, but have a higher $K_m$ for $H_2$.

As normal Fe-S clusters are $n = 1$ redox groups and hydrogenases react with an $n = 2$ substrate ($H_2$), these enzymes may be depicted as composed of two parts (Fig. 1.). Part 1 contains the $H_2$-activating site (Ni or a 6Fe cluster) and is involved in the transfer of two reducing equivalents at a time. Part II forms the contact of the enzyme with the metabolism of the microorganism and usually reacts with $n = 1$ redox components like ferredoxin or cytochromes. In some enzymes a two-electron donor/acceptor reacts here, e.g., $NAD^+$ or $F_{420}$, and then part II contains flavin as well.

[1] University of Amsterdam, E.C. Slater Institute for Biochemical Research, Plantage Muidergracht 12, NL-1018 TV Amsterdam, The Netherlands

41. Colloquium Mosbach 1990
The Molecular Basis of Bacterial Metabolism

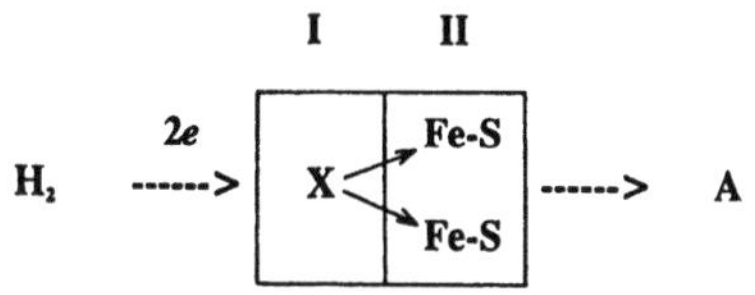


**Fig. 1.** Schematic representation of hydrogenases

## 2 Nickel Hydrogenases

Purified hydrogenases from many bacteria contain stoichiometric amounts of EPR-detectable, redox-active Ni, as first established for the enzyme from *Methanobacterium thermoautotrophicum*, strain Marburg (Graf and Thauer 1981; Albracht et al. 1982). Since that time the developments in the field have been reviewed several times (Odom and Peck 1984; Walsh and Orme-Johnson 1987; Hausinger 1987; Cammack 1988; Lancaster 1988; Fauque et al. 1988).

The majority of nickel hydrogenases are stable towards oxygen, although this gas completely prevents the reaction with $H_2$. Carbon monoxide is a competitive inhibitor, except possibly in hydrogenases from *Knallgas* bacteria.

### *2.1 Redox Properties of Nickel and Enzyme Activity*

In most aerobic solutions of Ni hydrogenases, the nickel ion is trivalent with very similar EPR signatures: one or two rhombic signals within the same preparation with g values between g = 2 and g = 2.3 (Fig. 2, upper two traces). The major difference in the two rhombic EPR signals is the position of the $g_y$ lines: 2.24 or 2.16. Both signals are due to low-spin Ni(III), a $3d^7$ system, in a rhombically-distorted octahedral ligand field, as first proposed by Lancaster (1980).

In 1985 Fernandez et al. (1985) demonstrated that the activity of enzyme preparations from *Desulfovibrio gigas* was proportional to the Ni(III) signal with $g_y$ = 2.16 ("ready" state). Enzyme molecules with Ni(III) having $g_y$ = 2.24 were not active. They were "unready" to react with $H_2$, but could be fully activated

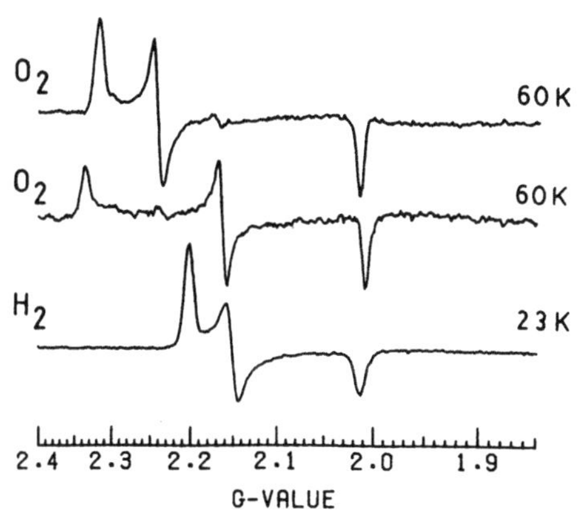


**Fig. 2.** EPR signatures of Ni(III) (*upper* two spectra) and Ni-C (*lower* spectrum) in *C. vinosum* hydrogenase

after incubation under reducing conditions for several hours. In this chapter, nickel in "ready" and "unready" enzymes will be referred to as $Ni_r$ and $Ni_u$, respectively.

Redox-titrations with the *D. gigas* enzyme, at room temperature and in the presence of mediators, only reveal reversible titrations of $Ni_u$(III) [$E'_0$ (pH 7) = −130 mV, n = 1, −60 mV per pH unit; Cammack et al. 1982; Cammack 1988; Moura et al. 1982; Teixeira et al. 1989]. Although $Ni_r$(III) could be reduced, it did not titrate in a reversible way. Similar observations were made with hydrogenases from *M. thermoautotrophicum* and *Chromatium vinosum* at pH 6 (Coremans et al. 1989; J.M.C.C. Coremans , unpubl.).

An aerobic enzyme has to undergo two activation stages before the "active" state is formed (Fernandez et al. 1985, 1986; Teixeira et al. 1985; Coremans et al. 1989). Oxygen has to be removed (deoxygenation) and the enzyme has to be incubated for some time under reducing conditions (reductive activation). Nickel in "active" enzymes will be referred to as $Ni_{act}$. In the "active" state, a third transient EPR signal from Ni can be observed, as was first demonstrated by Moura et al. (1982) in the *D. gigas* enzyme. It is often called Ni-C (Fig. 2, lower trace).

The events leading to "active" nickel in hydrogenases may be summarized as in Fig. 3. Reaction 1 is only reversible for $Ni_u$. The formation of $Ni_r$(II) in reaction 2 is a very slow process. The formation of $Ni_{act}$(II) in reaction 3 is a rapid process. Reactions 2 and 3 are both part of the reductive activation. It is not understood which redox components, if any, are involved (Lissolo et al. 1984; Fernandez et al. 1984). Reactions 4 and 5 are reversible (n = 1) in the presence of mediators. Only Ni-C is EPR detectable; its valence state is a matter of dispute. As the midpoint potentials of reactions 4 and 5 are rather close, at most 50% of the Ni can be observed as Ni-C. The character of "$Ni_{act}$(0)" is not understood. The $Ni_{act}$(II) in active enzymes is *not* in redox equilibrium with $Ni_r$(III) or $Ni_u$(III) (Coremans et al. 1989; J.M.C.C. Coremans, unpubl.). No lag phase in activity is observed in oxidized enzymes lacking a Ni(III) signal: they might be in the $Ni_{act}$(II) or $Ni_r$(II) state.

### 2.2 *The Ni-C State: Light Sensitivity and Binding of $H_2$ and CO*

Both $H_2$ and CO can bind to Ni-C in axial position; the bond is broken upon illumination (Van der Zwaan et al. 1985, 1986, 1988, 1990). The evidence is summarized in Fig. 4. With $^{13}CO$, a nearly isotropic superhyperfine splitting (3 mT) from one $^{13}C$ (I = 1/2) nucleus was detected. Hydrogen is presumably bound as $H_2$ rather than an $H^-$ ion or $H^+$. In model compounds of Ni(I)-H, a nearly isotropic protonhyperfine splitting of 10 mT has been observed (Morton and Preston 1984). In Ni-C an interaction of only a few tenths of a mT has been observed. Taken the extremely low $H_2$-partial pressures at which some microorganisms can still thrive (down to $2.10^{-8}$ bar; Cord-Ruwisch et al. 1988) it cannot be excluded that also $Ni_{act}$(II) can bind and activate $H_2$ in a kinetically competent way. The actual splitting of the $H_2$ molecule is a heterolytic process (Krasna 1979): $EH_2 \leftrightarrow EH^- + H^+$. As the Ni-C signal is also observed at room temperature (M. Chen and S.P.J. Albracht, unpubl.), the equilibrium of this reaction lies far to the left.

In inactive enzyme:

1. $Ni_{u,r}$ (III) → $Ni_{u,r}$(II)
2. $Ni_u$(II) → $Ni_r$(II)
3. $Ni_r$(II) → $Ni_{act}$(II)

In active enzyme:

4. $Ni_{act}$ (II) → Ni-C {$Ni_{act}$(I)}
5. Ni-C → '$Ni_{act}$ (0)'

**Fig. 3.** Redox changes of nickel and conformational changes of the enzyme in inactive (unready and ready) and active Ni hydrogenases

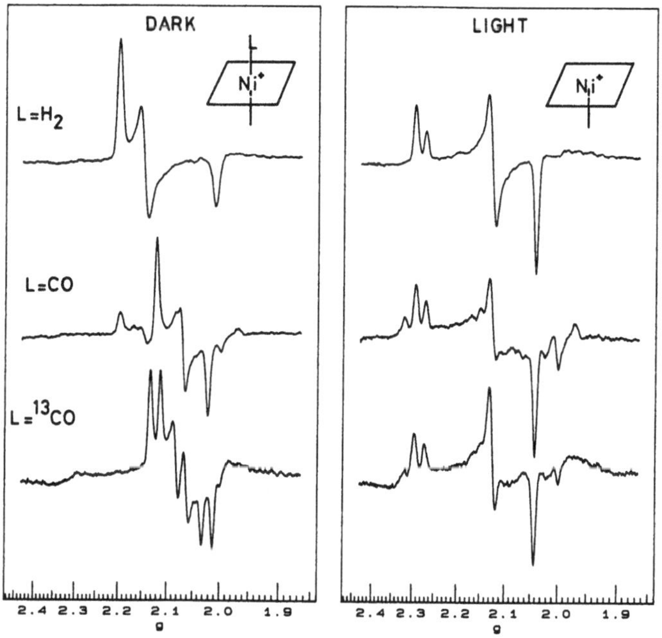


**Fig. 4.** Binding of $H_2$, CO and $^{13}CO$ to Ni-C in hydrogenase from *C. vinosum* (EPR spectra, *left*), and the effect of light (at 25 K) thereupon (*right*). After photodissociation, essentially the same spectrum was obtained in all cases. Only the photodissociation of the nickel-$H_2$ bond was retarded (5.6 times) if $^2H_2/^2H_2O$ was used (after Van der Zwaan et al. 1990)

### *2.3 Formal Valence State of Ni-C*

Van der Zwaan et al. (1986, 1990) favored monovalent nickel in a rhombically distorted octahedral environment with $H_2$ bound in axial position, to understand the properties of Ni-C and its oxidation to Ni(III). Other investigators assume Ni-C to be trivalent with a bound $H^-$ ion (Teixeira et al. 1989) or with bound dihydrogen (Cammack et al. 1988). The latter two states are isoelectronic, as is a Ni(I).$H^+$ state.

## 2.4 Sulphur and Nitrogen as Ligands to Nickel

There are indications for a weakly interacting N nucleus, possibly due to a distal N from histidine, in some hydrogenases (Tan et al. 1984; Chapman et al. 1988; Cammack et al. 1989). EXAFS measurements indicated the presence of three S ligands (Lindahl et al. 1984) or more (Scott et al. 1984, 1985). From studies on the effect of $^{33}S$ ($I = 3/2$) enrichment of hydrogenase from *Wolinella succinogenes* on the EPR spectra, Albracht et al. (1986) concluded that one to two S atoms were ligands to Ni. On basis of the interpretation of Ni-C as Ni(I)-$H_2$, Van der Zwaan et al. (1990) argued that N is less likely to be a ligand to Ni. It was proposed that three equatorial O atoms and one equatorial S atom, in addition to one axial S atom, form the ligands to Ni in Ni-S hydrogenases.

## 2.5 Selenium as a Ligand to Nickel

Selenium has now been found in Ni hydrogenases from five to six different bacteria. It is a direct ligand to Ni in the periplasmic *D. baculatus* enzyme (Eidsness et al. 1989; one Se at 2.44 Å and one to two S atoms at 2.17 Å). EPR with $^{77}Se$-enriched ($I = 1/2$) reduced enzyme ($g_{x,y,z} = 2.228, 2.174, 2.01$; He et al. 1989) showed a considerable superhyperfine effect of $^{77}Se$ ($A_{x,y} = 1.0$ mT, 1.8 mT). Regrettably, the possible light sensitivity of this state of the enzyme was not tested. It may be argued that the superhyperfine interaction in the x-y region, in combination with the finding of one g value close to $g_e$, indicates that Se is probably in axial position.

A TGA codon for selenocysteine is present in the gene encoding for the large subunit of the periplasmic *D. baculatus* enzyme. The position of this codon is very similar to a TGC codon for cysteine found in the gene encoding the large subunit of the Ni-S hydrogenase of *D. gigas* (He et al. 1989). The presence of Se is proposed to influences the kinetics of binding and exchange of the hydride ion at the Ni site (Fauque et al. 1988).

## 2.6 Possible Function of the [3Fe-4S] Cluster

Not all Ni hydrogenases contain a [3Fe-4S] cluster, so it is not required for activity measured in the standard assays. In their natural habitat purple sulphur bacteria, like *Thiocapsa roseopersicina* and *C. vinosum*, are daily exposed to changing growth conditions: from photolithotrophic under anaerobic conditions during the day-time, to chemolithotrophic under semiaerobic conditions at night. They can quickly change their energy metabolism from one mode to the other (Kämpf and Pfennig 1986; De Wit and Van Gemerden 1987). Hydrogenase in these bacteria must be able to respond to these daily changes. It is the author's impression that a [3Fe-4S] cluster might be of help here because of the following:

1. Its presence might prevent the formation of the dangerous superoxide radical:

$$Ni(II) + O_2 \rightarrow Ni(III) + O_2^-$$
$$Ni(II) + [3Fe\text{-}4S]^0 + O_2 \rightarrow Ni(III) + [3Fe\text{-}4S]^+ + O_2^{2-}$$

Purified $H_2$-reduced *C. vinosum* hydrogenase does not irreversibly inactivate when mixed with air. In contrast, the $F_{420}$-nonreducing enzyme from *M. thermoautotrophicum*, which lacks a [3Fe-4S] cluster (Albracht et al. 1982; Johnson et al. 1986), easily denatures after this treatment, as also apparent from a manifold of protein bands on SDS-gels (R.D. Fontijn and S.P.J. Albracht, unpubl.). The damage might be caused by $O_2^-$ radicals produced close to the nickel centre.

2. In deoxygenated oxidized enzymes, containing a $[3Fe\text{-}4S]^+$ cluster, the trivalent nickel in either the 'unready' or 'ready' state and the 3Fe cluster can be reduced by $H_2$ within minutes (Fernandez et al. 1986; Van der Zwaan pers. commun.):

$$H_2 + Ni(III) + [3Fe\text{-}4S]^+ \rightarrow Ni(II) + [3Fe\text{-}4S]^0 + 2H^+$$

Apparently both $Ni_u(III)$ and $Ni_r(III)$ can bind and activate $H_2$ with slow rates. In the $F_{420}$-nonreducing hydrogenase from *M. thermoautotrophicum*, reduction of Ni(III) by $H_2$ is extremely slow unless redox mediators are added (Coremans pers. commun.):

$$H_2 + Ni(III) + A \rightarrow Ni(II) + A^- + 2H^+$$

### 2.7 *Nature of the 4Fe Clusters*

Mössbauer spectroscopy of the *D. gigas* enzyme clearly shows the presence of two $[4Fe\text{-}4S]^{2+(2+;1+)}$ clusters, be it with some atypical magnetic properties (Teixeira et al. 1989). Other properties likewise distinguish them from the well-characterized "bacterial" 4Fe cluster. They have unusually broad and featureless EPR spectra which bear no resemblance with those from classical cubane clusters ('g = 1.94' type). In the amino-acid sequence of Ni hydrogenases no stretches typical for normal cubane clusters can be found (Fauque et al. 1988; Leclerc et al. 1988). The precise coordination of these clusters thus remains to be established.

## 3 Iron Hydrogenases

Six different bacteria are now known to contain Fe hydrogenases. The best characterized enzymes are those from *Clostridium pasteurianum* and *D. vulgaris*. Iron hydrogenases are usually extremely sensitive towards oxygen and are purified under strictly anaerobic conditions.

The confusion on the number of Fe atoms in these enzymes, reported to range from 4 to 12 (Chen and Mortenson 1974; Erbes et al. 1975; Adams and Mortenson 1984) in the *C. pasteurianum* hydrogenase I, has considerably retarded the progress in the understanding of the enzymes. In 1984, Chen and Blanchard, working with the *C. pasteurianum* hydrogenase II, noted that the calculated Fe content of this enzyme was greatly dependent on the method of protein determination. The group of Veeger further investigated this problem. After establishing the amino-acid sequence of the *D. vulgaris* hydrogenase from the encoding genes (Voordouw et al. 1985; Voordouw and Brenner 1985), this knowledge was used to calculate the number of Fe atoms, i.e., 15.3 (Hagen et al. 1986a,b). Hagen et al. noted that some of the conventional methods for protein determination gave erroneous results with this enzyme. As the mature small subunit was 34 amino-acid residues smaller than encoded by the gene (Prickril et al. 1986), the Fe content was adjusted to 14.4 and the acid-labile sulphur content to 12.3 (Filipiak et al. 1987). The number of 14 Fe atoms per molecule of enzyme (MW 55029) led Hagen et al. to postulate the presence of a novel 6Fe cluster as the active site, in addition to two classical $[4Fe\text{-}4S]^{2+(2+;1+)}$ clusters.

Two amino-acid sequence patterns, containing cysteins, and characteristic for [4Fe-4S] proteins (Cys-X-X-Cys-X-X-Cys-X-X-X-Cys-Pro) are present in the N-terminal region of the 46-kDa subunit (Voordouw and Brenner 1985). The remaining 10 cysteine residues are not grouped in patterns normally observed in Fe-S proteins. The mature small subunit contains no cysteine (Prickril et al. 1986). The two structural genes have been expressed in *Escherichia coli*, resulting in an inactive form of the enzyme (Voordouw et al. 1987). Although this protein contained the two [4Fe-4S] clusters, it lacked all physicochemical properties ascribed to the novel 6Fe cluster.

The Fe content in the *C. pasteurianum* enzymes was recently reevaluated (Adams et al. 1989). The Type-I enzyme (20.1 Fe atoms) is now considered to contain four cubane clusters and one 6Fe cluster. The Type-II enzyme (13.8 Fe atoms) contains two classical cubane clusters and one 6Fe cluster (called Hydrogen-activating cluster or H-cluster). Both types of clusters have clearly different Mössbauer characteristics (Rusnak et al. 1987).

Fifteen years ago, Erbes et al. (1975) already performed EPR studies with the *C. pasteurianum* hydrogenase I to test whether $H_2$ or its competitor CO might bind to the Fe-S clusters (Fig. 5). With the current knowledge on Fe-S proteins one can easily see that the spectrum of the reduced enzyme looks like that of two interacting $[4Fe\text{-}4S]^{1+}$ clusters, but it could not be interpreted so at that time since only 4Fe and 4 $S^{2-}$ atoms were calculated to be present in the enzyme. In view of the recent findings (Adams et al. 1989), the Fe content in this enzyme was presumably at least twice as high. When reoxidized with 2,6-dichlorophenolindophenol (DCIP) or $H^+$ (removal of $H_2$), a rhombic signal with all g values greater than 2 (2.099, 2.041, 2.001) was observed with a spin concentration of 0.07–0.13 spins per 4 Fe. Misled by this apparently low paramagnetism in the oxidized state, Erbes et al. (1975) suspected that "this state was artifactual or at least not an obligatory state of the enzyme during turnover". In the same article, however, these investigators reported the first, and until now only, pre-steady-state experiments with hydrogenase. The rhombic signal in the oxidized state completely disappeared within 6 ms after mixing with $H_2$, whereby 75% of the signal of the reduced enzyme appeared. When starting with $H_2$-reduced enzyme (75 $\mu$M $H_2$, pH 8) and mixing with 200 $\mu$M DCIP, the "reduced" signal disappeared within 6 ms and the rhombic "oxidized" signal had already achieved 60% of its maximum value. The authors concluded that the species responsible for both signals were kinetically competent to play a role in the catalytic cycle. These early observations are highly relevant to current ideas on the active site.

Erbes et al. (1975) also described that $D_2/D_2O$ had no effect on the line widths of the EPR spectra of the oxidized enzyme. The effect of 0.1 bar of CO on the

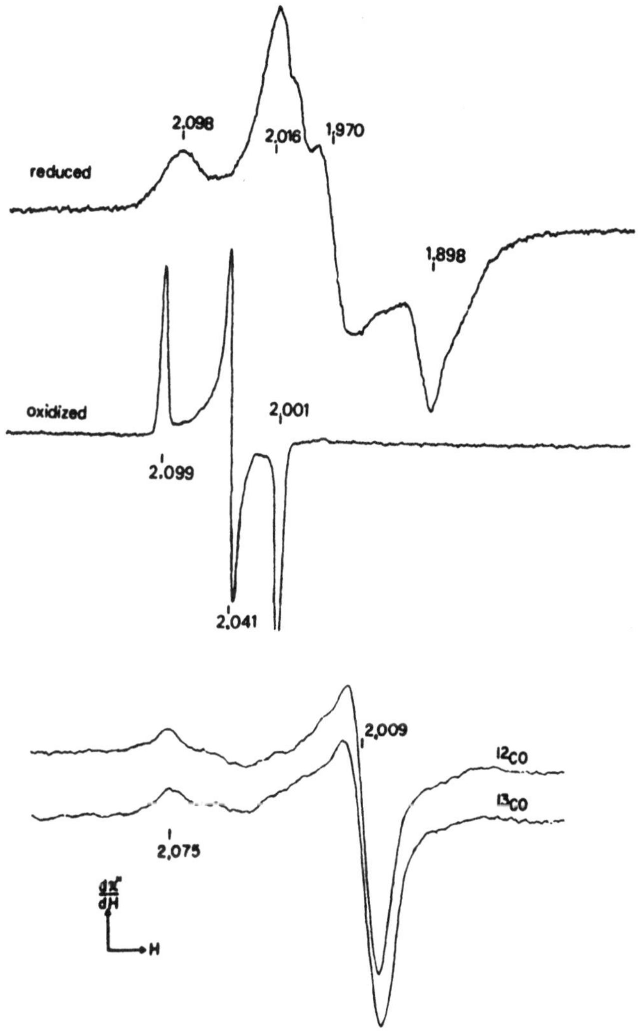


**Fig. 5.** EPR spectra of hydrogenase I from *C. pasteurianum* in the reduced (*upper* trace), oxidized (*middle* trace) and oxidized CO-treated form ($^{12}CO$ or $^{13}CO$; *lower* two traces; Erbes et al. 1975)

"oxidized" spectrum was dramatic: it completely disappeared and an axial signal ($g_{x,y}$ = 2.009, $g_z$ = 2.075) appeared (Fig. 5, lower traces; reversible after removal of CO). Carbon monoxide 90% enriched in $^{13}C$ (I = 1/2) gave a very small broadening. When combined with current knowledge, it can be concluded that this was the first evidence that the H-cluster binds the competitive inhibitor CO.

Very similar rhombic and axial signals have now also been detected in the enzymes from *M.elsdenii* (Van Dijk et al. 1980), *D. vulgaris* (Hagen et al. 1986a; Patil et al. 1988a) and the Type-II enzyme from *C. pasteurianum* (Zambrano et al. 1989). The midpoint potential of the H-cluster is low: –400 mV (pH 8) in the *C.*

*pasteurianum* enzymes (Adams 1987) and –320 mV (pH 7) in the *D. vulgaris* enzyme (Patil et al. 1988a).

It has been long known that the CO inhibition of $H_2$ formation in growing cells of *C. butyricum* (Kempner and Kubowitz 1933), or of a crude extract of *C. pasteurianum* (Thauer et al. 1974), could be reversed by illumination. Patil et al. (1986, 1988b) demonstrated that illumination at 9 K of the *D. vulgaris* enzyme with the CO-induced axial signal, resulted in a slow conversion of the axial signal into the rhombic one (reversible after warming to 200 K for 5 min).

Also $O_2$ can bind to the H-cluster, as demonstrated with the *C. pasteurianum* hydrogenase I (Kowal et al. 1988). Just like CO, low concentrations of $O_2$ could convert the rhombic signal into the axial one. In both cases, illumination at 8 K caused the transition from the axial signal into the rhombic one. Contact of $O_2$ with active Fe hydrogenases leads to inactivation.

There is evidence for N coordination to the H-cluster in the *C. pasteurianum* II enzyme (George et al. 1989). With EXAFS (George et al. 1989) the average Fe-Fe distance in the oxidized enzyme was found to be 2.8 Å, but in reduced enzyme an additional Fe-Fe distance of 3.3 Å, ascribed to the 6Fe cluster was observed. This Fe-Fe distance is similar to Fe-O-Fe in hemerythrins. In none of the known 6Fe model clusters is the Fe-Fe distance equal to 3.3 Å (George et al. 1989).

It may be concluded that the species responsible for the rhombic 2.10 signal (6Fe cluster) probably forms the primary site of interaction of $H_2$ and CO (and $O_2$) with the enzyme (Fig. 6).

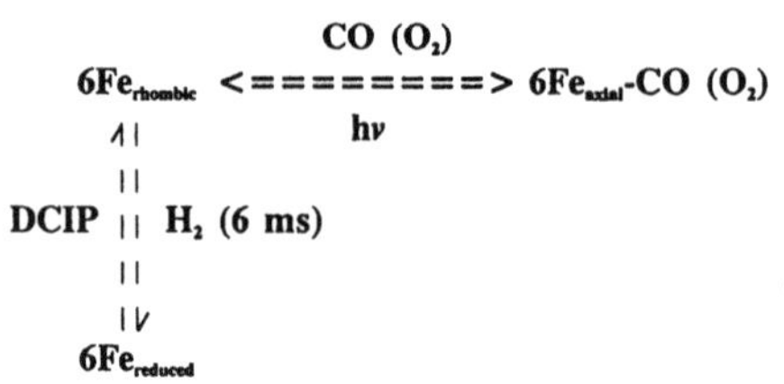


**Fig. 6.** Summary of reactions of the H-clusters in Fe hydrogenases

## References

Adams, M.W.W. (1987) The mechanism of $H_2$ activation and CO binding by hydrogenase I and hydrogenase II of *Clostridium pasteurianum*. J. Biol. Chem. 262:15054–15061

Adams, M.W.W. & Mortenson, L.E. (1984) The purification of hydrogenase II (uptake hydrogenase) from the anaerobic $N_2$-fixing bacterium *Clostridium pasteurianum*. Biochim. Biophys. Acta 766:51–61

Adams, M.W.W., Mortenson, L.E. & Chen, J.-S. (1980) Hydrogenase. Biochim. Biophys. Acta 594:105–176

Adams, M.W.W., Eccleston, E. & Howard, J.B. (1989) Iron-sulfur clusters of hydrogenase I and hydrogenase II of *Clostridium pasteurianum*. Proc. Natl. Acad. Sci. USA 86:4932–4936

Albracht, S.P.J., Graf, E.-G. & Thauer, R.K. (1982) The EPR properties of nickel in hydrogenase from *Methanobacterium thermoautotrophicum*. FEBS Lett. 140:311–313

Albracht, S.P.J., Kröger, A., Van der Zwaan, J.W., Unden, G., Böcher, R., Mell, H. & Fontijn, R.D. (1986) Direct evidence for sulphur as a ligand to nickel in hydrogenase: an EPR study of the enzyme from *Wolinella succinogenes* enriched in $^{33}S$. Biochim. Biophys. Acta 874:116–127

Cammack, R. (1988) Nickel in metalloproteins. Adv. Inorg. Chem. 32:297–333

Cammack, R., Patil, D.S., Aguirre, R. & Hatchikian, E.C. (1982) Redox properties of the ESR-detectable nickel in hydrogenase from *Desulfovibrio gigas*. FEBS Lett. 142:289–292
Cammack, R., Chapman, A. & Hatchikian, E.C. (1988) Solvent accessibility of metal centres in a nickel-hydrogenase. 2nd Int. Symp. Mol. Biol. Hydrogenase, Helen, Georgia, USA (Abstr)
Cammack, R., Kovacs, K.L., McCracken, J. & Peisach, J. (1989) Spectroscopic characterization of the nickel and iron-sulphur clusters of hydrogenase from the purple photosynthetic bacterium *Thiocapsa roseopersicina*. 2. Electron spin-echo spectroscopy. Eur. J. Biochem. 182:363–366
Chapman, A., Cammack, R., Hatchikian, E.C., McCracken, J. & Peisach, J. (1988) A pulsed EPR study of redox-dependent hyperfine interactions for the nickel centre of *Desulfovibrio gigas* hydrogenase. FEBS Lett. 242:134–138
Chen, J.-S. & Blanchard, D.K. (1984) Purification and properties of the $H_2$-oxidizing (uptake) hydrogenase of the $N_2$-fixing anaerobe *Clostridium pasterianum* W5. Biochem. Biophys. Res. Commun. 122:9–16
Chen, J.-S. & Mortenson, L.E. (1974) Purification and properties of hydrogenase from *Clostridium pasteurianum* W5. Biochim. Biophys. Acta 371:283–298
Cord-Ruwisch, R., Seitz, H.-J. & Conrad, R. (1988) The capacity of hydrogenotrophic anaerobic bacteria to compete for traces of hydrogen depends on the redox potential of the terminal electron acceptor. Arch. Microbiol. 149:350–357
Coremans, J.M.C.C., Van der Zwaan, J.W. & Albracht, S.P.J. (1989) Redox behaviour of nickel in hydrogenase from *Methanobacterium thermoautotrophicum* (strain Marburg). Correlation between the nickel valence state and enzyme activity. Biochim. Biophys. Acta 997:256–267
De Wit, R. & Van Gemerden, H. (1987) Chemolithotrophic growth of the phototrophic sulfur bacterium *Thiocapsa roseopersicina*. FEMS Microbiol. Ecol. 45:117–126
Eidsness, M.K., Scott, R.A., Prickril, B.C., DerVartanian, D.V., LeGall, J., Moura, I., Moura, J.J.G. & Peck, H.D., Jr. (1989) Evidence for selenocysteine coordination to the active site nickel in the [NiFeSe] hydrogenase from *Desulfovibrio baculatus*. Proc. Natl. Acad. Sci. USA 86:147–151
Erbes, D.L., Burris, R.H. & Orme-Johnson, W.H. (1975) On the iron-sulfur cluster in hydrogenase from *Clostridium pasteurianum* W5. Proc. Natl. Acad. Sci. USA 72:4795–4799
Fauque, G., Peck, H.D., Jr., Moura, J.J.G., Huynh, B.H., Berlier, Y., DerVartanian, D.V., Teixeira, M., Przybyla, A.E. & Lespinat, P.A. (1988) The three classes of hydrogenases from sulfate-reducing bacteria of the genus *Desulfovibrio*. FEMS Microbiol. Rev. 54:299–344
Fernandez, V.M., Aguirre, R. & Hatchikian, E.C. (1984) Reductive activation and redox properties of hydrogenase from *Desulfovibrio gigas*. Biochim. Biophys. Acta 790:1–7
Fernandez, V.M., Hatchikian, E.C. & Cammack, R. (1985) Properties and reactivation of two different deactivated forms of *Desulfovibrio gigas* hydrogenase. Biochim. Biophys. Acta 832:69–79
Fernandez, V.M., Hatchikian, E.C., Patil, D.S. & Cammack, R. (1986) ESR-detectable nickel and iron-sulphur centres in relation to the reversible activation of *Desulfovibrio gigas* hydrogenase. Biochim. Biophys. Acta 883:145–154
Filipiak, M., Hagen, W.R., Grande, H.J., Dunham, W.R., Van Berkel-Arts, A., Krüse-Wolters, M. & Veeger, C. (1987) Structural and magnetic properties of Fe-hydrogenases reinvestigated. Rec. Trav. Chim. Pays Bas 106:230
George, G.N., Prince, R.C., Stockley, K.E. & Adams, M.W.W. (1989) X-ray absorption-spectroscopic evidence for a novel iron cluster in hydrogenase II from *Clostridium pasteurianum*. Biochem. J. 259:597–600
Graf, E.-G. & Thauer, R.K. (1981) Hydrogenase from *Methanobacterium thermoautotrophicum*: a nickel-containing enzyme. FEBS Lett. 136:165–169
Hagen, W.R., Van Berkel-Arts, A., Krüse-Wolters, K.M., Dunham, W.R. & Veeger, C. (1986a) EPR of a novel high-spin component in activated hydrogenase from *Desulfovibrio vulgaris* (Hildenborough). FEBS Lett. 201:158–162
Hagen, W.R., Van Berkel-Arts, A., Krüse-Wolters, K.M., Voordouw, G. & Veeger, C. (1986b) The iron-sulfur composition of the active site of hydrogenase from *Desulfovibrio vulgaris* (Hildenborough) deduced from its subunit structure and total iron-sulfur content. FEBS Lett. 203:59–63
Hausinger, R.P. (1987) Nickel utilization by microorganisms. Microbiol. Rev. 51:22–42
He, S.H., Teixeira, M., LeGall, J., Patil, D.S., Moura, I., Moura, J.J.G., DerVartanian, D.V., Huynh, B.H. & Peck, H.D., Jr. (1989) EPR studies with $^{77}$Se-enriched (NiFeSe) hydrogenase of *Desulfovibrio baculatus*. Evidence for a selenium ligand to the active site nickel. J. Biol. Chem. 264:2678–2682

Johnson, M.K., Zambrano, I.C., Czechowski, M.H., Peck, H.D., Jr., DerVartanian, D.V. & LeGall, J. (1986) Magnetic circular dichroism and electron paramagnetic resonance studies of nickel-containing hydrogenases. In: Xavier, A.V. (ed) Frontiers in bioinorganic chemistry. VCH Verlagsgesellschaft mbH, Weinheim, FRG, pp. 36–44

Kämpf, C. & Pfennig, N. (1986) Chemoautotrophic growth of *Thiocystis violacea, Chromatium gracile* and *C. vinosum* in the dark at various $O_2$-concentrations. J. Basic Microbiol. 26:517–531

Kempner, W. & Kubowitz, F. (1933) Wirkung des Lichtes auf die Kohlenoxydhemmung der Buttersäuregärung. Biochem. Z. 265:245–252

Kowal, A.T., Adams, M.W.W. & Johnson, M.K. (1988) Electron paramagnetic resonance studies of the low temperature photolytic behavior of oxidized hydrogenase I from *Clostridium pasteurianum*. 2nd Int. Symp. Mol. Biol. Hydrogenase, Helen, Georgia, USA (Abstr)

Krasna, A.I. (1979) Hydrogenase: properties and applications. Enzyme Microb. Technol. 1:165–172

Lancaster, J.R., Jr. (1980) Soluble and membrane-bound paramagnetic centers in *Methanobacterium bryantii*. FEBS Lett. 115:285–288

Lancaster, J.R., Jr. (1988) The bioinorganic chemistry of nickel. VCH Verlagsgesellschaft mbH, Weinheim, FRG

Leclerc, M., Colbeau, A., Cauvin, B. & Vignais, P.M. (1988) Cloning and sequencing of the genes encoding the large and the small subunits of the $H_2$ uptake hydrogenase (hup) of *Rhodobacter capsulatus*. Mol. Gen. Genet. 214:97–107

Li, C., Peck, H.D., Jr., LeGall, J. & Przybyla, A.E. (1987) Cloning, characterization, and sequencing of the genes encoding the large and small subunits of the periplasmic (NiFe) hydrogenase of *Desulfovibrio gigas*. DNA 6:539–551

Lindahl, P.A., Kojima, N., Hausinger, R.P., Fox, J.A., Teo, B.K., Walsh, C.T. & Orme-Johnson, W.H. (1984) Nickel and iron EXAFS of $F_{420}$-reducing hydrogenase from *Methanobacterium thermoautotrophicum*. J. Am. Chem. Soc. 106:3062–3064

Lissolo, T., Pulvin, S. & Thomas, D. (1984) Reactivation of the hydrogenase from *Desulfovibrio gigas* by hydrogen. Influence of redox potential. J. Biol. Chem. 259:11725–11729

Lissolo, T., Choi, E.S., LeGall, J. & Peck, H.D., Jr. (1986) The presence of multiple intrinsic membrane nickel-containing hydrogenases in *Desulfovibrio vulgaris* (Hildenborough). Biochem. Biophys. Res. Commun. 139:701–708

Morton, J.R. & Preston, K.F. (1984) An ESR study at 4 K of the reaction between H and $Ni(CO)_4$. J. Chem. Phys. 81:5775–5778

Moura, J.J.G., Moura, I., Huynh, B.H., Krüger, H.-J., Teixeira, M., DuVarney, R.G., DerVartanian, D.V., Xavier, A.V., Peck, H.D., Jr. & LeGall, J. (1982) Unambiguous identification of the nickel EPR signal in $^{61}Ni$-enriched *Desulfovibrio gigas* hydrogenase. Biochem. Biophys. Res. Commun. 108:1388–1393

Odom, J.M. & Peck, H.D., Jr. (1981) Hydrogen cycling as a general mechanism for energy coupling in the sulfate-reducing bacteria, *Desulfovibrio* sp. FEMS Microbiol. Lett. 12:47–50

Odom, J.M. & Peck, H.D., Jr. (1984) Hydrogenase, electron-transfer proteins and energy coupling in the sulfate-reducing bacteria *Desulfovibrio*. Ann. Rev. Microbiol. 38:551–592

Patil, D.S., Czechowski, M.H., Huynh, B.H., LeGall, J., Peck, H.D., Jr. & DerVartanian, D.V. (1986) A reversible effect of low carbon monoxide concentrations on the EPR spectra of the periplasmic hydrogenase from *Desulfovibrio vulgaris*. Biochem. Biophys. Res. Commun. 137:1086–1093

Patil, D.S., Moura, J.J.G., He, S.H., Teixeira, M., Prickril, B.C., DerVartanian, D.V., Peck, H.D., Jr., LeGall, J. & Huynh B.H. (1988a) EPR-detectable redox centers of the periplasmic hydrogenase from *Desulfovibrio vulgaris*. J. Biol. Chem. 263:18732–18738

Patil, D.S., Huynh, B.H., He, S.H., Peck, H.D., Jr., DerVartanian, D.V. & LeGall, J. (1988b) A reversible photoconversion between the carbon monoxide induced axial 2.06 and the rhombic 2.10 EPR signals of the periplasmic hydrogenase from *Desulfovibrio vulgaris*. J. Am. Chem. Soc. 110:8533–8534

Prickril, B.C., Czechowski, M.H., Przybyla, A.E., Peck, H.D., Jr. & LeGall, J. (1986) A putative signal peptide on the smaller subunit of the periplasmic hydrogenase from *Desulfovibrio vulgaris* (Hildenborough). J. Bacteriol. 167:722–725

Rusnak, F.M., Adams, M.W.W., Mortenson, L.E. & Münck E. (1987) Mössbauer study of *Clostridium pasteurianum* hydrogenase II: evidence for a novel three-iron cluster. J. Biol. Chem. 262:38–41

Scott, R.A., Wallin, S.A., Czechowski, M., DerVartanian, D.V., LeGall, J., Peck, H.D., Jr. & Moura, I.

(1984) X-ray absorption spectroscopy of nickel in the hydrogenase from *Desulfovibrio gigas*. J. Am. Chem. Soc. 106:6864–6865
Scott, R.A., Czechowski, M., DerVartanian, D.V., LeGall, J., Peck, H.D., Jr. & Moura, I. (1985) Nickel X-ray absorption spectroscopy of *Desulfovibrio gigas* hydrogenase. Rev. Port. Quim. 27:67–70
Tan, S.-L., Fox, J.A., Kojima, N., Walsh, C.T. & Orme-Johnson, W.H. (1984) Nickel coordination in deazoflavin and viologen-reducing hydrogenase from *Methanobacterium thermoautotrophicum*. J. Am. Chem. Soc. 106:3064–3066
Teixeira, M., Moura, I., Xavier, A.V., Huynh, B.H., DerVartanian, D.V., Peck, H.D., Jr., LeGall, J. & Moura, J.J.G. (1985) EPR studies on the mechanism of activation and the catalytic cycle of the nickel-containing hydrogenase from *Desulfovibrio gigas*. J. Biol. Chem. 260:8942–8950
Teixeira, M., Moura, I., Xavier, A.V., Moura, J.J.G., LeGall, J., DerVartanian, D.V., Peck, H.D., Jr. & Huynh, B.H. (1989) Redox intermediates of *Desulfovibrio gigas* [NiFe] hydrogenase generated under hydrogen. Mössbauer and EPR characterization of the metal centers. J. Biol. Chem. 264:16435–16450
Thauer, R.K., Käufer, B., Zähringer, M. & Jungermann, K. (1974) The reaction of the iron-sulfur protein hydrogenase with carbon monoxide. Eur. J. Biochem. 42:447–452
Van der Zwaan, J.W., Albracht, S.P.J., Fontijn, R.D. & Slater, E.C. (1985) Monovalent nickel in hydrogenase from *Chromatium vinosum*. FEBS Lett. 179:271–277
Van der Zwaan, J.W., Albracht, S.P.J., Fontijn, R.D.& Roelofs, Y.B.M. (1986) EPR evidence for direct interaction of carbon monoxide with nickel in hydrogenase from *Chromatium vinosum*. Biochim. Biophys. Acta 872:208–215
Van der Zwaan, J.W., Bouwens, E.C.M. & Albracht, S.P.J. (1988) Probing the active site of hydrogenase from *Chromatium vinosum*. 2nd Int. Symp. Mol. Biol. Hydrogenase, Helen, Georgia, USA (Abstr)
Van der Zwaan, J.W., Coremans, J.M.C.C., Bouwens, E.C.M. & Albracht, S.P.J. (1990) Effect of $^{17}O_2$ and $^{13}CO$ on EPR spectra of nickel in hydrogenase from *Chromatium vinosum*. Biochim. Biophys. Acta (in press)
Van Dijk, C., Grande, H.J., Mayhew, S.G. & Veeger, C. (1980) Properties of hydrogenase from *Megasphaera elsdenii*. Eur. J. Biochem. 107:251–261
Voordouw, G. & Brenner, S. (1985) Nucleotide sequence of the gene encoding the hydrogenase from *Desulfovibrio vulgaris* (Hildenborough). Eur. J. Biochem. 148:515–520
Voordouw, G., Walker, J.E. & Brenner, S. (1985) Cloning of the gene encoding for the hydrogenase from *Desulfovibrio vulgaris* (Hildenborough) and determination of the $NH_2$-terminal sequence. Eur. J. Biochem. 148:509–514
Voordouw, G., Hagen, W.R., Krüse-Wolters, K.M., Van Berkel-Arts, A. & Veeger, C. (1987) Purification and characterisation of *Desulfovibrio vulgaris* (Hildenborough) hydrogenase expressed in *Eschericia coli*. Eur. J. Biochem. 162:31–36
Walsh, C.T. & Orme-Johnson, W.H. (1987) Nickel enzymes. Biochemistry 26:4901–4906
Zambrano, I.C., Kowal, A.T., Mortenson, L.E., Adams, M.W.W. & Johnson, M.K. (1989) Magnetic circular dichroism and electron magnetic paramagnetism studies of hydrogenase I and II from *Clostridium pasteurianum* J. Biol. Chem. 264:20974–20983

# Three Nitrogen Fixation Systems in *Azotobacter vinelandii*

P.E. Bishop, R.D. Joerger, and R. Premakumar[1]

## 1 Introduction

In the early 1980s we presented evidence indicating that the $N_2$-fixing bacterium, *Azotobacter vinelandii*, contained at least two nitrogenase systems: the conventional Mo-containing nitrogenase (nitrogenase-1) system; and an alternative nitrogenase system expressed in the absence of Mo. This evidence primarily centered on the observation that $Nif^-$ (unable to fix $N_2$) mutant strains underwent phenotypic reversal (i.e., $Nif^-$ to $Nif^+$) under conditions of Mo deprivation. These reports (Bishop et al. 1980; Bishop et al. 1982; Page and Collinson 1982; Premakumar et al. 1984) were received with skepticism because they challenged the long-held belief that Mo was absolutely required for $N_2$ fixation and that nitrogenases were essentially the same regardless of their source. The latter notion was further supported by the results of Southern blot experiments by Ruvkun and Ausubel (1980) which indicated that some of the structural genes encoding nitrogenases from diverse diazotrophic organisms were highly conserved at the nucleotide sequence level.

Since the $Nif^-$ strains of *A. vinelandii* that were first used to demonstrate phenotypic reversal probably contained point mutations, it was considered possible that phenotypic reversal was due to increased leakiness of the mutant phenotypes under conditions of Mo starvation, and not to derepression of an alternative $N_2$-fixation system. This possibility was ruled out when mutant strains containing deletions in the structural genes for nitrogenase-1 were observed to exhibit phenotypic reversal under Mo-deficient conditions (Bishop et al. 1986a,b; Robson 1986).

*A. vinelandii* has now been shown to harbor two alternative Mo-independent nitrogenases: a V-containing enzyme complex (nitrogenase-2); and a nitrogenase (nitrogenase-3) that does not appear to contain either Mo or V (Hales et al. 1986a,b; Chisnell et al. 1988).

## 2 Nitrogenase-1: A Mo-Containing Nitrogenase

Nitrogenase-1 is only expressed in medium containing molybdenum (Mo) and consists of two components: dinitrogenase reductase-1 (also called component II or Fe protein); and dinitrogenase-1 (also designated component I or MoFe

[1]Department of Microbiology and United States Department of Agriculture, Agricultural Research Service, North Carolina State University, Raleigh, North Carolina 27695–7615, USA

41. Colloquium Mosbach 1990
The Molecular Basis of Bacterial Metabolism

protein). Dinitrogenase reductase-1, which serves as an electron donor to dinitrogenase-1, is a dimer of two identical subunits with an $M_r$ of approximately 60 000. A single [4Fe-4S] cluster is bridged between the two subunits (Gillum et al. 1977; Hausinger and Howard 1983). Dinitrogenase-1, with an $M_r$ of about 220 000, is a tetramer that is made up of two pairs of nonidentical subunits ($\alpha$ and $\beta$). Dinitrogenase-1 contains two types of metal centers that are involved in the redox reactions of the $N_2$ reduction process: P centers that might be organized as four unusual 4Fe-4S clusters (Eady 1986; McLean et al. 1988); and two identical FeMo cofactors (FeMoco), which are almost certainly the sites for $N_2$ binding and reduction (Smith et al. 1985).

## 3 Mo-Independent Nitrogenases

### *3.1 Nitrogenase-2*

Nitrogenase-2 is a vanadium-containing enzyme complex which is synthesized in N-free medium lacking Mo but containing vanadium (V) (Hales et al. 1986a,b; Robson et al. 1986a). This enzyme complex consists of two components: dinitrogenase reductase-2, a dimer of two identical subunits; and dinitrogenase-2, which is now thought to be a hexamer ($M_r$ of about 240 000) of two dissimilar pairs of large subunits ($\alpha$ and $\beta$) and a pair of small subunits ($\delta$) (Robson et al. 1989). Dinitrogenase reductase-2 has an $M_r$ of about 62 000 and contains four Fe atoms and four acid-labile sulfide groups per dimer (Eady et al. 1988; Hales et al. 1986b). Dinitrogenase-2 contains 2 V atoms, 23 Fe atoms and 20 acid-labile sulfide groups per molecule (Eady et al. 1987). A cofactor (FeVaco) analogous to FeMoco has been extracted from dinitrogenase-2 using N-methylformamide (Smith et al. 1988).

### *3.2 Nitrogenase-3*

Nitrogenase-3 does not appear to contain either Mo or V and is made under Mo- and V-deficient conditions (Chisnell et al. 1988). This enzyme is composed of two components, dinitrogenase reductase-3 and dinitrogenase-3. Dinitrogenase reductase-3 is a dimer ($M_r$ of approximately 65 000) of two identical subunits, while dinitrogenase-3 is a tetramer ($M_r$ of about 216 000) composed of two dissimilar pairs of subunits ($\alpha$ and $\beta$). However, dinitrogenase-3 may actually be a hexamer, since the structural gene operon for nitrogenase-3 contains an open reading frame (ORF) that potentially encodes a protein which is similar to the $\delta$ subunit for dinitrogenase-2 (Joerger et al. 1989b). Dinitrogenase reductase-3 contains four Fe atoms and four acid-labile sulfide groups per dimer. Dinitrogenase-3 contains approximately 24 Fe atoms and 18 acid-labile sulfide groups per molecule; however, it lacks significant amounts of Mo or V.

## 4 *nif* Genes Shared by the Three Nitrogenase Systems in *Azotobacter vinelandii*

Two *nif* genes are known to be shared by the three nitrogenase systems. One of these is *nifB*, since NifB⁻ mutants (strains CA30 and UW45) are unable to grow under Mo-sufficient, Mo-deficient plus V, and Mo-deficient diazotrophic conditions (Joerger and Bishop 1988; Joerger et al. 1986). In the wild type, a *nifB*-hybridizing transcript (4 kb in size) is also observed under all three diazotrophic conditions (Joerger and Bishop 1988). Since NifB is known to be required for FeMoco synthesis, it is probably safe to assume that this gene product plays an analogous role in the formation of FeVaco [the V-containing cofactor of dinitrogenase-2 (Smith et al. 1988)]. In the case of dinitrogenase-3, NifB must be involved in some function other than one that relates to Mo or V. The other *nif* gene shared by the three nitrogenase systems is *nifM*. NifM is required for maturation of dinitrogenase reductase-1 and a NifM⁻ mutant (strain MV21) was unable to grow diazotrophically either in the presence or absence of Mo or V (Kennedy et al. 1986).

## 5 Genes Encoding Nitrogenase-2

The structural genes encoding dinitrogenase-2 and dinitrogenase reductase-2 have been cloned, sequenced, and mutagenized for both *A. chroococcum* (Robson et al. 1986b; Robson et al. 1989) and *A. vinelandii* (Joerger et al. 1990; Raina et al. 1988). These genes have been designated as *vnf* (*v*anadium *n*itrogen *f*ixation; Robson et al. 1989). In contrast to the single operon (*nifHDK*) encoding the subunits for nitrogenase-1, the genes encoding the nitrogenase-2 proteins are split between two operons (Fig. 1). *vnfH* encodes the dinitrogenase reductase-2 subunits and is part of a two-gene operon. The ORF 3′ to *vnfH* encodes a ferredoxinlike protein which has not been ascribed a function. *vnfH* is preceded by a potential promoter sequence that would be predicted to interact with core RNA polymerase containing the sigma 54 factor (*ntrA*, *rpoN*, or *glnF* gene product). The *vnfDGK* operon, located 1.0 kbp (*A. vinelandii*) or 2.5 kbp (*A. chroococcum*) downstream from the *vnfH-Fd* operon, encodes the subunits for dinitrogenase-2. *vnfD* encodes the $\alpha$-subunit and *vnfK* encodes the $\beta$-subunit. In *A. vinelandii* the 1.0-kbp region between the *vnfH-Fd* and *vnfDGK* operons does not appear to contain any identifiable ORFs (Joerger et al. 1990).

The third subunit, $\delta$, ($M_r$ of 13,274) is encoded by *vnfG* (Robson et al. 1989). This gene does not have a counterpart in the nitrogenase-1 system (system 1 in Fig. 1); however, it does share some sequence similarity to *anfG*, an ORF located between *anfD* and *anfK* (Fig. 1). Whether or not the $\delta$ subunit is required for full activity of dinitrogenase-2 is presently unknown. The *vnfDGK* genes appear to be cotranscribed and transcription is $NH_4^+$ repressible. The largest and most abundant transcript is 3.4 kb in length and the two transcripts present in lesser amounts are 1.9 and 1.7 kb in size (Robson et al. 1989).

The removal of a 1.4-kbp *Bgl*II fragment, which spans all of *vnfG* and the 3′ and 5′ ends of *vnfD* and *vnfK*, from the genomes of both *A. chroococcum* (Robson et al. 1989) and *A. vinelandii* (Pau et al. 1989) result in deletion strains that lack dinitrogenase-2. When this deletion was transferred to *A. chroococcum* strain

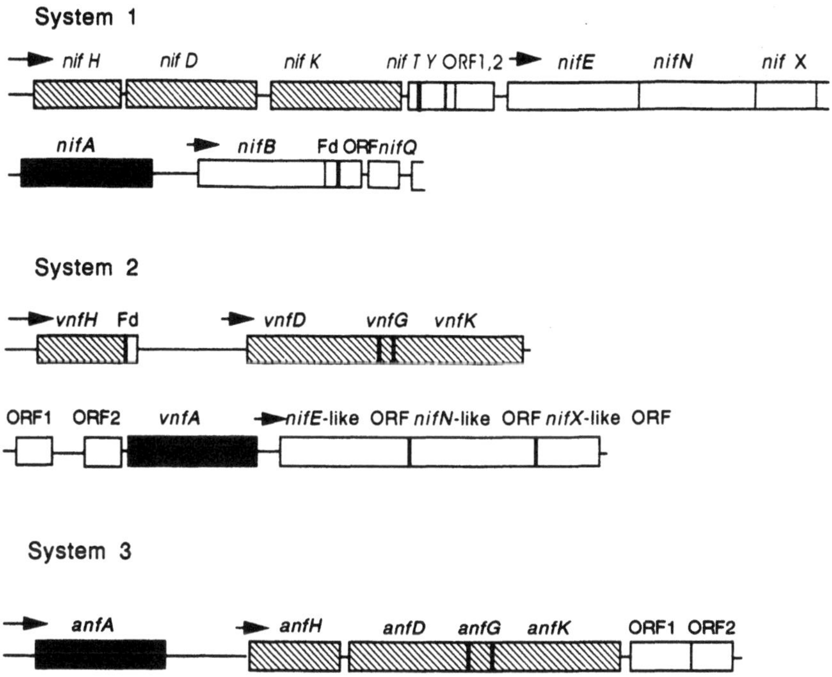


**Fig. 1.** Organization of some genes involved in the three nitrogenase systems of *A. vinelandii*. Systems *1, 2* and *3* refer to genes required for nitrogenases 1, 2, and 3, respectively. *Arrows* indicate promoters and direction of transcription

MCD1155 (carrying a deletion of the structural genes for nitrogenase-1), the resulting double deletion strain was unable to grow under any $N_2$-fixing condition. This result indicates that *A. chroococcum* does not contain a third nitrogenase (Robson et al. 1989). A similar double deletion strain of *A. vinelandii* (strain RP206) grew in N-free media lacking Mo (Pau et al. 1989). This finding provided genetic evidence for the expression of a third nitrogenase (nitrogenase-3) in *A. vinelandii* which lacks Mo and V, and also supported previously published results describing the isolation and partial characterization of nitrogenase-3 (Chisnell et al. 1988).

## 6 Genes Encoding Nitrogenase-3

The structural genes encoding nitrogenase-3 in *A. vinelandii* have been cloned, sequenced, and mutagenized (Joerger et al. 1989b). These genes have been designated *anf* (*a*lternative *n*itrogen *f*ixation) and they are organized in a single operon, *anfHDGK*, *ORF1*, *ORF2* (Fig. 1). The subunits of dinitrogenase reductase-3 are encoded by *anfH*, while the $\alpha$ and $\beta$ subunits of dinitrogenase-3 are encoded by *anfD* and *anfK*, respectively. A third subunit ($\delta$) for dinitrogenase-3 is probably encoded by *anfG*. The *anfHDGK* operon is preceded by a potential promoter sequence that would interact with RNA polymerase containing sigma 54.

The predicted protein products of the two ORFs that are located 3′ to *anfK* (Fig. 1) do not show overall similarity to any known *nif* gene products. However, the predicted ORF1 product contains some sequence identity to the $NH_2$-terminal part of dinitrogenase reductase and another region exhibits identity to presumed heme-binding domains of P-450 cytochromes. The predicted product of ORF2 does not seem to show any interesting similarity with other amino acid sequences in the Bionet data base (Joerger et al. 1989b). Deletions plus insertions placed in several regions of the *anfHDGK* operon resulted in Anf⁻ mutants that were unable to grow in N-free, Mo-deficient medium. However, growth in media containing Mo or V was normal (Joerger et al. 1989b). The absence of nitrogenase-3 proteins in these Anf⁻ mutants was also confirmed by two-dimensional gel electrophoresis (Joerger et al. 1989b).

## 7 Nonstructural Genes

Recently two *nifA*-like ORFs were identified in DNA cloned from *A. vinelandii*. One of these ORFs was recognized in the DNA sequence flanking the Tn*5* insertion carried by a Vnf⁻ mutant (strain CA46) that is unable to express nitrogenase-2 when derepressed in N-free medium containing V (Joerger et al. 1989a; Joerger et al. 1986). Since this mutant synthesizes both nitrogenases-1 and -3 under Mo-deficient, N-free conditions in the presence or absence of V, the *nifA*-like ORF was designated *vnfA* (Joerger et al. 1989a). The other *nifA*-like ORF was located approximately 700 bp upstream from *anfH* (Fig. 1). A mutant (strain CA66), carrying a deletion plus insertion in this *nifA*-like ORF, synthesized only nitrogenase-2 proteins after derepression in Mo-deficient media with or without V. Thus this ORF was designated *anfA* (Joerger et al. 1989a). The highest degree of similarity between the predicted products of *nifA*, *vnfA*, and *anfA* is in the C-terminal half of the proteins, where the potential RNA polymerase-sigma 54 interaction sites and ATP-binding domains are located. A potential DNA-binding domain (Fisher et al. 1988) that includes a helix-turn-helix motif is present in both predicted products of *vnfA* and *anfA* (Joerger et al. 1989a).

*nifENX*-like ORFs are located immediately downstream from *vnfA* (Wolfinger et al. 1989). Preliminary results with a strain containing Tn*5*-*lacZ* inserted towards the 3′ end of the *nifN*-like ORF indicate that the *nifN*- and/or *nifX*-like ORF is required for diazotrophic growth in Mo-deficient media with or without V. However, in the presence of Mo this strain shows wild-type growth (E. Wolfinger and P. Bishop, unpubl. results). These results imply that these *nifENX*-like genes are required for functional nitrogenases-2 and -3 but not nitrogenase-1.

## 8 Sequence Comparisons Between Structural Genes

In Table 1, sequence comparisons are shown for the structural genes and their presumed protein products. Overall sequence identity is greater for *nifH*, *vnfH*, and *anfH* and their products (dinitrogenase reductases) than for the genes encoding the

**Table 1.** Sequence comparisons between *nif*, *vnf*, and *anf* genes of *A. vinelandii*[a]

| Gene comparisons | Percent nucleotide sequence identity | Percent amino acid sequence identity of predicted gene products |
|---|---|---|
| *nifH* × *vnfH* | 88.5 | 91.0 |
| *nifH* × *anfH* | 69.3 | 62.8 |
| *vnfH* × *anfH* | 70.1 | 63.5 |
| *nifD* × *vnfD* | 52.6 | 33.0 |
| *nifD* × *anfD* | 49.6 | 32.7 |
| *vnfD* × *anfD* | 65.8 | 54.4 |
| *vnfG* × *anfG* | 55.9 | 39.8 |
| *nifK* × *vnfK* | 51.5 | 31.1 |
| *nifK* × *anfK* | 50.7 | 32.1 |
| *vnfK* × *anfK* | 69.8 | 57.4 |
| *nifD* × *nifK* | – | 19.8 |
| *vnfD* × *vnfK* | – | 24.5 |
| *anfD* × *anfK* | – | 21.5 |

[a] The values shown in this table were derived from published sequence data (Jacobson et al. 1989; Joerger et al. 1989b; Joerger et al. 1990).

three dinitrogenases. Five cysteine residues are conserved across the three dinitrogenase reductase proteins, and two of these conserved residues appear to serve as ligands for the Fe-S center which is bound symmetrically between the dinitrogenase reductase-1 subunits (Joerger et al. 1989b). A motif characteristic of nucleotide-binding domains (Gly-X-Gly-XX-Gly) is also present in the three gene products (Joerger et al. 1989b). The percent identity between *nifH* and *vnfH* is quite high (88.5%) and suggests that these two genes may have diverged relatively recently (in evolutionary time) from an ancestral gene, or that functional constraints are very similar for the two dinitrogenase reductases. In contrast, *anfH* seems to be more distantly related to *nifH* and *vnfH* (about 70% identity). This correlates well with the inability of dinitrogenase reductase-3 to yield high nitrogenase activity in a complementation assay with dinitrogenase-1. In contrast, dinitrogenase reductase-2 gives high activity in the same assay (Chisnell et al. 1988).

The *vnfDK* and *anfDK* genes are more similar to each other than either set of genes is to *nifDK*, with *vnfK* and *anfK* sharing slightly more identity (69.8%) than *vnfD* and *anfD* (65.8%). The products of *nifDK*, *vnfDK* and *anfDK* contain cysteine and histidine residues that are conserved in nearly all dinitrogenase proteins which have thus far been examined. These highly conserved Cys and His residues are thought to be coordinating ligands for Fe-S centers. Some of these Cys residues are essential for dinitrogenase activity as demonstrated by site-directed mutagenesis experiments (Brigle et al. 1987).

As previously mentioned, *vnfG* and *anfG* show identity with respect to both nucleotide sequence (55.9%) and amino acid sequences of the presumed products

(39.8%). This may indicate that these gene products function as subunits of their respective dinitrogenases in a similar fashion.

Finally, the percent identity between amino acid sequences of the $\alpha$ and $\beta$ subunits of each dinitrogenase (Table 1) suggests that the genes encoding these two subunits may have evolved from a common ancestral gene, as previously suggested for Mo-containing dinitrogenases (Thöny et al. 1985). Based on these identity comparisons, it has been speculated that *nifD* and *nifK* may have diverged somewhat earlier during evolution than the genes encoding the subunits of dinitrogenases-2 and -3 (Joerger et al. 1989b; Robson et al. 1989).

An observation that may relate to the occurrence of alternative nitrogenases in other organisms is the finding that predicted products of *nifH*-like genes from two very different diazotrophs exhibit a high degree of identity with the *anfH* product (Table 2). One of these is NifH3 (81.7% identity) from the obligate anaerobe, *Clostridium pasteurianum*, and the other is NifH1 (72.4% identity) from the thermophilic archaebacterium, *Methanococcus thermolithothrophicus*. It remains to be seen, however, whether or not these identities signify functional similarity to the nitrogenase-3 system. In the case of *nifH3* from *C. pasteurianum*, no evidence for transcription could be found under Mo-sufficient diazotrophic conditions; however, transcription under Mo-deficient conditions was not examined (Wang et al. 1988). Although the predicted product of *nifH1* from *M. thermolithotrophicus* shows a fairly high degree of identity (72.4%) with the *anfH* product, it is clear that the *M. thermolithotrophicus nifD* product shows much less identity (38.3%) with the *anfD* product; therefore it can be concluded that the degree of identity between amino acid sequences of dinitrogenase reductase proteins does not necessarily

**Table 2.** Sequence comparisons of predicted products of *nif*, *vnf*, and *anf* genes from *A. vinelandii* and of *nif* genes from *Methanococcus thermolithotrophicus* and *Clostridium pasteurianum*[a]

| Gene product comparison | Percent amino acid identity |
|---|---|
| Av NifH × Cp NifH3 | 63.2 |
| Av NifH × Mt NifH1 | 63.4 |
| Av VnfH × Cp NifH3 | 62.3 |
| Av VnfH × Mt NifH1 | 63.0 |
| Av AnfH × Cp NifH3 | 81.7 |
| Av AnfH × Mt NifH1 | 72.4 |
| Cp NifH3 × Mt NifH1 | 70.3 |
| Av NifD × Mt NifD | 40.4 |
| Av VnfD × Mt NifD | 41.5 |
| Av AnfD × Mt NifD | 38.3 |

[a] Source of sequence: *A. vinelandii*, AV (see footnote to Table 1); *Clostridium pasteurianum*. Cp (Wang et al. 1988); and *Methanococcus thermolithotrophicus*, Mt (Souillard and Sibold 1989).

correlate with the percent identity observed for the $\alpha$ and $\beta$ subunits of different dinitrogenases (also see Table 1). This higher degree of variability is also observed with the conventional Mo-containing dinitrogenases (Eady 1986).

## References


Bishop, P.E., Jarlenski, D.M.L. & Hetherington, D.R. (1980) Evidence for an alternative nitogen fixation system in *Azotobacter vinelandii*. Proc. Natl. Acad. Sci. USA 77:7342–7346

Bishop, P.E., Jarlenski, D.M.L. & Hetherington, D.R. (1982) Expression of an alternative nitrogen fixation system in *Azotobacter vinelandii*. J. Bacteriol. 150:1244–1251

Bishop, P.E., Hawkins, M.E. & Eady, R.R. (1986a) Nitrogen fixation in Mo-deficient continuous culture by a strain of *Azotobacter vinelandii* carrying a deletion of the structural genes for nitrogenase (*nifHDK*). Biochem. J. 238:437–442

Bishop, P.E., Premakumar, R., Dean, D.R., Jacobson, M.R., Chisnell, J.R., Rizzo, T.M. & Kopczynski, J. (1986b) Nitrogen fixation by *Azotobacter vinelandii* strains having deletions in structural genes for nitrogenase. Science 232:92–94

Brigle, K.E., Setterquist, R.A., Dean, D.R., Cantwell, J.S., Weiss, M.C. & Newton, W.E. (1987) Site-directed mutagenesis of the nitrogenase MoFe protein of *Azotobacter vinelandii*. Proc. Natl. Acad. Sci. USA 84:7066–7069

Chisnell, J.R., Premakumar, R. & Bishop, P.E. (1988) Purification of a second alternative nitrogenase from a *nifHDK* deletion strain of *Azotobacter vinelandii*. J. Bacteriol. 170:27–33

Eady, R.R. (1986) Enzymology in free-living diazotrophs. In: Broughton, W.J., Pühler, A. (eds.) Nitrogen fixation. Oxford, Clarendon Press 4:1–49

Eady, R.R., Robson, R.L., Richardson, T.H., Miller, R.W. & Hawkins, M. (1987) The vanadium nitrogenase of *Azotobacter chroococcum*: purification and properties of the VFe protein. Biochem. J. 244:197–207

Eady, R.R., Richardson, T.H., Miller, R.W., Hawkins, M. & Lowe, D.J. (1988) The vanadium nitrogenase of *Azotobacter chroococcum*: purification and properties of the Fe protein. Biochem. J. 256:189–196

Fischer, M.-H., Bruderer, T. & Hennecke, H. (1988) Essential and nonessential domains in the *Bradyrhizobium japonicum* NifA protein: identification of indispensable cysteine residues involved in redox reactivity and/or metal binding. Nucleic Acids Res. 16:2207–2224

Gillum, W.O., Mortenson, L.E. Chen, J.S. & Holm, R.H. (1977) Quantitative extrusion of the $Fe_4S_4$ cores of active sites of ferredoxins and the hydrogenase of *Clostridium pasteurianum*. J. Am. Chem. Soc. 99:584–595

Hales, B.J., Case, E.E., Morningstar, J.E., Dzeda, M.F. & Mauterer, L.A., (1986a) Isolation of a new vanadium-containing nitrogenase from *Azotobacter vinelandii*. Biochemistry 25:7251–7255

Hales, B.J., Langosch, D.J. & Case, E.E. (1986b) Isolation and characterization of a second nitrogenase Fe-protein from *Azotobacter vinelandii*. J. Biol. Chem. 261:15301–15306

Hausinger, R.P. & Howard, J.B. (1983) Thiol reactivity of the nitrogenase Fe-protein from *Azotobacter vinelandii*. J. Biol. Chem. 258:13486–13492

Jacobson, M.R., Brigle, K.E., Bennett, L.T., Setterquist, R.A., Wilson, M.S., Cash, V.L., Beynon, J., Newton, W.E. & Dean, D.R. (1989) Physical and genetic map of the major *nif* gene cluster from *Azotobacter vinelandii*. J. Bacteriol. 171:1017–1027

Joerger, R.D. & Bishop, P.E. (1988) Nucleotide sequence and genetic analysis of the *nifB-nifQ* region from *Azotobacter vinelandii*. J. Bacteriol. 170:1475–1487

Joerger, R.D., Premakumar, R. & Bishop, P.E. (1986) Tn*5*-induced mutants of *Azotobacter vinelandii* affected in nitrogen fixation under Mo-deficient and Mo-sufficient conditions. J. Bacteriol. 168:673–682

Joerger, R.D., Jacobson, M.R. & Bishop, P.E. (1989a) Two *nifA*-like genes required for expression of alternative nitrogenases by *Azotobacter vinelandii*. J. Bacteriol. 171:3258–3267

Joerger, R.D., Jacobson, M.R., Premakumar, R., Wolfinger, E.D. & Bishop, P.E. (1989b) Nucleotide sequence and mutational analysis of the structural genes (*anfHDGK*) for the second alternative nitrogenase from *Azotobacter vinelandii*. J. Bacteriol. 171:1075–1086

Joerger, R.D., Loveless, T.M., Pau, R.N., Mitchenall, L.A., Simon, B.H. & Bishop, P.E. (1990) Nucleotide sequences and mutational analysis of the structural genes for nitrogenase 2 of *Azotobacter vinelandii*. J. Bacteriol. 172:3400–3408

Kennedy, C., Gamal, R., Humphrey, R., Ramos, J., Brigle, K. & Dean, D. (1986) The *nifH*, and *nifN* genes of *Azotobacter vinelandii*: characterization by Tn*5* mutagenesis and isolation from pLAFR1 gene banks. Mol. Gen. Genet. 205:318–325

McLean, P.A., Papaefthymiou, V., Münck, E. & Orme-Johnson, W.H. (1988) Use of isotopic hybrids of the MoFe protein to study the mechanism of nitrogenase catalysis. In: Bothe, H., deBruijn, F.J., Newton, W.E. (eds.) Nitrogen fixation: hundred years after. Fischer, Stuttgart, pp. 101–106

Page, W.J. & Collinson, S.K. (1982) Molybdenum enhancement of nitrogen fixation in a Mo-starved *Azotobacter vinelandii* $Nif^-$ mutant. Can. J. Microbiol. 28:1173–1180

Pau, R.N., Mitchenall, L.A. & Robson, R.L. (1989) Genetic evidence for an *Azotobacter vinelandii* nitrogenase lacking molybdenum and vanadium. J. Bacteriol. 171:124–129

Premakumar, R., Lemos, E.M. & Bishop, P.E. (1984) Evidence for two dinitrogenase reductases under regulatory control by molybdenum in *Azotobacter vinelandii*. Biochim. Biophys. Acta 797:64–70

Raina, R., Reddy, M.A., Ghosal, D. & Das, H.K. (1988) Characterization of the gene for the Fe-protein of the vanadium dependent alternative nitrogenase of *Azotobacter vinelandii* and construction of a Tn*5* mutant. Mol. Gen. Genet. 214:121–127

Robson, R.L. (1986) Nitrogen fixation in strains of *Azotobacter chroococcum* bearing deletions of a cluster of genes coding for nitrogenase. Arch. Microbiol. 146:74–79

Robson, R.L., Eady, R.R., Richardson, T.H., Miller, R.W., Hawkins, M. & Postgate, J.R. (1986a) The alternative nitrogenase of *Azotobacter chroococcum* is a vanadium enzyme. Nature (Lond) 322:388–390

Robson, R.L., Woodley, P.R. & Jones, R. (1986b) Second gene (*nifH**) coding for a nitrogenase iron-protein in *Azotobacter chroococcum* is adjacent to a gene coding for a ferredoxin-like protein. EMBO J. 5:1159–1163

Robson, R.L., Woodley, P.R., Pau, R.N. & Eady, R.R. (1989) Structural genes for the vanadium nitrogenase from *Azotobacter chroococcum*. EMBO J. 8:1217–1224

Ruvkun, G.B. & Ausubel, F.M. (1980) Interspecies homology of nitrogenase genes. Proc. Natl. Acad. Sci. USA 77:191–195

Smith, B.E., Bishop, P.E., Dixon, R.A., Eady, R.R., Filler, W.A., Lowe, D.J., Richards, A.J.M., Thomson, A.J., Thornley, R.N.F. & Postgate, J.R. (1985) The iron-molybdenum cofactor of nitrogenase. In: Evans, H.J., Bottomley, P.J., Newton, W.E. (eds.) Nitrogen fixation research progress. Nijhoff, Dordrecht Boston, pp. 597–603

Smith, B.E., Eady, R.R., Lowe, D.J. & Gormal, C. (1988) The vanadium-iron protein of vanadium nitrogenase from *Azotobacter chroococcum* contains an iron-vanadium cofactor. Biochem. J. 250:299–302

Souillard, N. & Sibold, L. (1989) Primary structure, functional organization and expression of nitrogenase structural genes of the thermophilic archaebacterium *Methanococcus thermolithotrophicus*. Mol. Microbiol. 3:441–552

Thöny, B., Kaluza, K. & Hennecke, H. (1985) Structural and functional homology between the $\alpha$ and $\beta$ subunits of the nitrogenase MoFe protein as revealed by sequencing the *Rhizobium japonicum nifK* gene. Mol. Gen. Genet. 198:441–448

Wang, S.-Z., Chen, J.-S. & Johnson, J.L. (1988) The presence of five *nifH*-like sequences in *Clostridium pasteurianum*: sequence divergence and transcriptional properties. Nucl. Acids Res. 16:439–454

Wolfinger, E.D., Pau, R.N. & Bishop, P.E. (1989) Multiple *nifE*- and *nifN*-like genes in *Azotobacter vinelandii*. Ann. Mtg. Am. Soc. Microbiol. (Abstr H-97) 185

# From Nonsense to Sense: UGA Encodes Selenocysteine in Formate Dehydrogenase and Other Selenoproteins

A. BÖCK, C. BARON, K. FORCHHAMMER, J. HEIDER, W. LEINFELDER, G. SAWERS, B. VEPREK, E. ZEHELEIN, and F. ZINONI[1]

## 1 Introduction

Selenium is an element which resembles sulfur in its physical and chemical properties. In minerals, selenium occurs mainly in the form of selenides which are associated with the isomorphic sulfides in a ratio ranging between $1:10^3$ and $1:10^4$. The chemical forms in which selenium is available for plants or microorganisms, however, are the selenite and selenate anions. They are readily taken up by the plant root or the microbial cell and reduced intracellularly to the $Se^{2-}$ oxidation state. Little information is available concerning the details of these processes; it is still a matter of debate whether selenite and selenate uptake occurs via specific transport systems (Hudman and Glenn 1984), or whether the sulfate transporter is used (Lindblow-Kull et al. 1985). It is also still an open issue as to whether intracellular reduction of selenate and selenite is a purely chemical process, or whether it procedes via specific systems or via the sulfate reduction chain (for review see Brown and Shrift 1982; Doran 1982; Dürre and Andreesen 1986). The possibility that different organisms use different strategies also cannot be excluded.

The further metabolic fate of reduced selenium depends on the biological system and on the concentration in which selenium is offered; only microorganisms are considered in the following: (1) at low concentrations of selenite in the medium, usually 1 μM or less, selenium is covalently incorporated into proteins or into tRNA's by certain organisms (for review see Böck and Stadtman 1988). (2) at selenite concentrations grossly above 1 μM, bacterial sulfur metabolism is no longer able to discriminate between sulfur and selenium. As a consequence, selenium replaces sulfur in cellular components. This may result in the formation of iron sulfur centers in non-heme iron redox carriers in which S is partially replaced by Se (for review see Stadtman 1974) or in the synthesis of proteins in which methionine or cysteine residues are substituted by their selenoanalogues (see Stadtman et al. 1989 and references therein).

This review deals mainly with the specific incorporation of selenium into selenoproteins and seleno-tRNAs.

[1] Lehrstuhl für Mikrobiologie der Universität München, Maria-Ward-Straße la, W-8000 München 19, FRG

41. Colloquium Mosbach 1990
The Molecular Basis of Bacterial Metabolism

## 2 Results and Discussion

### 2.1 A UGA Codon Directs Insertion of Selenocysteine into Selenoproteins

Subunit A of glycine reductase from *Clostridium sticklandii* was the first protein for which it was shown that selenium is present in the form of selenocysteine (Cone et al. 1976). Other enzymes subsequently characterized as containing the selenocysteine moiety are listed in Table 1. Evidence for two of the listed enzymes is only indirect: the formate dehydrogenase $(FDH)_N$ from *E. coli* has been included in the list, since it could be shown that mutations which block selenocysteine biosynthesis and insertion into the 80-kDa selenopolypeptide subunit of $FDH_H$ also prevent formation of the 110-kDa subunit of $FDH_N$ (Leinfelder et al. 1988a). The hydrogenase of *Desulfovibrio baculatus* is included, since it was shown that the gene contains a TGA codon in a position where the genes for the large subunit of nonselenium hydrogenases contain a cysteine codon (Menon et al. 1988). All the enzymes listed in Table 1 share the characteristic of catalyzing oxidation reactions, which indicates that the -SeH group of the selenocysteine moiety may participate directly in catalysis.

In 1986, it was shown that the gene and the mRNA of glutathione peroxidase from mouse (Chambers et al. 1986), and the gene for the 80-kDa selenopolypeptide subunit of $FDH_H$ from *E. coli* (Zinoni et al. 1986) harbor an in-frame opal termination codon. In the case of the glutathione peroxidase gene, the position of the UGA codon corresponded to the position of the selenocysteine residue in the amino-acid sequence of the bovine enzyme (Günzler et al. 1984).

These findings prompted the postulation that selenocysteine insertion might occur cotranslationally. Direct proof that this is indeed the case was provided by gene fusion and localized mutagenesis experiments in the case of the *E. coli* $FDH_H$

**Table 1.** Selenocysteine-containing enzymes

| Enzyme | Organism | Size of selenopolypeptide | Reference |
|---|---|---|---|
| Glycine Reductase | *Clostridium sticklandii* | 12 kDa (A*/B/C)[a] | Cone et al. (1976) |
| Formate Dehydrogenase | H: *E. coli* | 80 kDa | Zinoni et al. (1986) |
| | N: *E. coli* | 110 kDa $(A_4{}^*/B_4/C_2)$ | Enoch and Lester (1975) |
| | *Clostridium thermoaceticum* | 100 kDa $(A_2{}^*/B_2)$ | Yamamoto et al. (1983) |
| | *Methanococcus vannielii* | 100 kDa $(A^*/B)_x$ | Jones and Stadtman (1981) |
| Hydrogenase | *Methanococcus vannielii* | 46 kDa $(A_2/B_4{}^*/C_2)$ | Yamazaki (1982) |
| | *Desulfovibrio baculatus* | 56 kDa (A/B*) | Vooourdow et al. (1989) |
| Glutathione peroxidase | Mammals | 21 kDa $(A_4{}^*)$ | Forstrom et al. (1978); Epp et al. (1983); Günzler et al. (1984) |

[a] The subunit structure (where known) is given in parentheses; the asterisk denotes the selenopeptide subunit.

selenopolypeptide (Zinoni et al. 1987). For glutathione peroxidase this still remains to be demonstrated. However, the fact that all the genes for the selenoproteins which have been sequenced until now contain an in-frame TGA codon [glutathione peroxidase genes from mouse (Chambers et al. 1986); man (Mullenbach et al. 1988); rat (Reddy et al. 1988); rabbit (Akasaka et al. 1989); FDH genes of *E. coli* (Zinoni et al. 1986); *Enterobacter aerogenes* (J. Heider, unpubl); and the hydrogenase gene of *D. baculatus* (Menon et al. 1988)] supports the contention that in both eubacteria and eukaryotes, selenocysteine is incorporated cotranslationally.

One of the most intriguing and also basic questions in this context is how the protein synthesis machinery discriminates between a UGA "nonsense" codon signaling polypeptide chain termination and a UGA "sense" codon directing the incorporation of selenocysteine. Theoretically, two requirements must be fulfilled: (1) there must be a tRNA species which recognizes the UGA "sense" codon; and (2) some peculiar feature of the mRNA must preclude that the polypeptide chain-release factor 2 (which is responsible for termination at UGA's) recognizes this particular codon. As described below, both requirements are fulfilled in *Escherichia coli*.

In an attempt to identify the features of the FDH mRNA which preclude termination at the UGA codon responsible for selenocysteine insertion, we constructed a series of deletions approaching the UGA codon from the 5′ and from the 3′ side. The gene cartidges obtained were cloned into the 5′ terminal portion of the *lacZ* gene and β-galactosidase formation was determined, which gave us a measure of translation of the UGA in the different cassettes. The surprising result was that a stretch of about 40 bases downstream of the UGA was absolutely necessary for efficient readthrough of the codon, and that a putative stem/loop structure could be formed within this segment. Deletions extending into the hairpin structure drastically reduced the decoding of the UGA by selenocysteine (F. Zinoni et al. 1990). It is still open as to whether it is the secondary structure per se or some specific sequence within it which is required for promotion of readthrough of the UGA and/or for prevention of termination at this position.

### 2.2 *Selenocysteine Biosynthesis*

To gather information on the path of selenocysteine synthesis and incorporation, mutants were isolated from *E. coli* which acidified the medium due to excretion of formate, through their inability to synthesize formate dehydrogenases. These mutants were analyzed for their ability to form the selenopolypeptide subunits of FDH and selenylated tRNAs (Leinfelder et al. 1988a). Similar pleiotropic FDH mutants (*fdh*) had been described in the literature but had not been correlated with any defect in selenium metabolism (Mandrand-Berthelot et al. 1978; Barrett et al. 1979). Several such *fdh* mutants were kindly provided by M.-A. Mandrand-Berthelot and included into the analysis. Four classes of mutants could be differentiated by genetic analysis. Their properties and the characteristics of the gene products are summarized in Table 2. Mutations in *selA*, *selB*, and *selC* block formation of selenoproteins but still allow the synthesis of selenylated tRNA's; mutations in *selD* prevent the incorporation of selenium in both macromolecules.

**Table 2.** Genes and their products involved in selenium metabolism in *E. coli*

| Gene | Map position (min) | Product | Function in synthesis of | | Reference |
|---|---|---|---|---|---|
| | | | Seleno-proteins | Seleno-tRNAs | |
| *selA* | 80 | 50 kDa protein | + | − | Leinfelder et al. (1988a) |
| *selB* | 80 | 70 kDa protein | + | − | Leinfelder et al. (1988a) |
| *selC* | 82 | tRNA$^{Ser}$ (UCA) | + | − | Leinfelder et al. (1988b) |
| *selD* | 38 | 37 kDa protein | + | + | Leinfelder et al. (1990) |

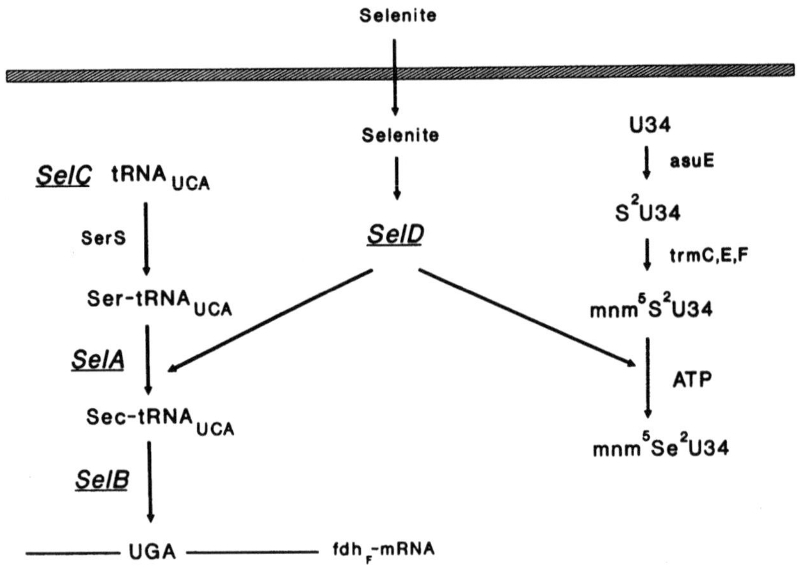


**Fig. 1.** Schematic representation of selenium metabolism in *E. coli*. Abbreviations: *SerS* Seryl-tRNA ligase; *U34* uridine residue at position 34 of tRNA; *S²U34* 2-thiouridine; *mnm⁵S²U34* 5-methyl-aminomethyl-2-thiouridine; *mnm⁵Se²U34* 5-methyl-aminomethyl-2-selenouridine; *fdh$_F$-mRNA* mRNA transcribed from the gene for the selenopolypeptide of formate dehydrogenase H from *E. coli*. *asuE* and *trmC,E,F* genes involved in modification of U34 according to Wittwer and Stadtman (1986)

Figure 1 depicts the steps of selenoprotein and seleno-tRNA synthesis blocked by the different mutations.

### *2.2.1 The selC Gene Product is a tRNA Species*

Determination of the nucleotide sequence of *selC* revealed that it codes for a tRNA species (Leinfelder et al. 1988b). This tRNA is highly unusual in that it possesses an 8 bp aminoacyl acceptor stem, an anticodon (UCA) which directly matches UGA, and deviating nucleotides in positions which hitherto were considered as being invariant in all tRNA molecules. The unusual features have been confirmed by directly sequencing the tRNA (Schön et al. 1989).

$tRNA^{UCA}$ is charged with L-serine by the seryl-tRNA ligase (Leinfelder et al. 1988b). Mutations in *selC* which abolished its capacity to accept L-serine concomitantly blocked its function in insertion of selenocysteine, thus strongly supporting the notion that selenocysteine biosynthesis occurs from an L-serine residue esterified to the tRNA (Fig. 1).

### *2.2.2 SelA and SelD are Involved in Conversion of Seryl-tRNA to Selenocysteinyl-tRNA*

To provide information on the biochemical role of the *selA, selB* and *selD* gene products, we have measured the ability of mutants with lesions in these genes to accumulate selenocysteinyl-tRNA in vivo. It was found that mutants with altered *selA* and *selD* genes were unable to form selenocysteinyl-tRNA, i.e., the respective gene products must have a function in the conversion of the seryl- into the selenocysteinyl-residue (Leinfelder et al. 1989). The SelA and the SelD proteins were purified using strains which overexpressed the respective genes (K. Forchhammer, B. Veprek and A. Böck, unpubl. results).

With the aid of the purified proteins, an in-vitro system was developed in which L-[$^{14}$C]serine charged to purified $tRNA^{UCA}$ was converted into tRNA-bound selenocysteine (Leinfelder et al. 1990). L-serine charged to cognate serine acceptor tRNA could not be converted to selenocysteine. Conversion required reduced selenium (either sodium selenide or selenite plus a thiol reagent) and $Mg^{2+} \times$ ATP.

A recent, more detailed analysis showed that SelA is the first enzyme to participate in this conversion reaction. SelA contains covalently bound pyridoxal-phosphate and dehydrates L-seryl-tRNA to the enzyme-bound, acrylyl-tRNA intermediate (K. Forchhammer and A. Böck, unpublished). SelD, in a second step, in the presence of $Mg^{2+} \times$ ATP and selenide formally catalyzes the 2,3 addition of $H_2Se$ to the acrylyl double bond resulting in selenocysteinyl-tRNA (B. Veprek and A. Böck, unpublished). The nature of the reactive selenium intermediate required for this reaction is still unknown. It is, however, interesting to note that the sulfur-selenium exchange occurring as the last step in synthesis of the selenylated nucleoside 5-methylaminomethyl-2-selenouridine (Wittwer and Stadtman 1986; see Fig. 1) also requires a functional *selD* gene product (Stadtman et al. 1989; Leinfelder et al. 1990). SelD, therefore, appears to function as a general donor of an activated selenium species. In contrast to recent reports (Mizutani et al. 1989), therefore, O-phosphorylation of L-serine does not take place in the selenocysteine formation sequence in *E. coli*.

### *2.2.3 The Product of the selB Gene is a Selenocysteinyl-tRNA-Specific Translation Factor*

*selB* mutants are defective in selenoprotein synthesis, although they are able to form selenocysteinyl-tRNA in vivo (Leinfelder et al. 1989). This result was corroborated by the finding that the in vitro conversion of seryl-tRNA into selenocysteinyl-tRNA does not require the presence of SelB (Leinfelder et al. 1990). This led to the conclusion that SelB may have a function in the decoding process itself, e.g., as a translation factor. The following lines of evidence strongly support this assumption: (1) SelB bears a high sequence similarity with the translation factors EF-Tu and IF2$\alpha$ over a stretch of 200 amino acids at the N-terminus; (2) SelB specifically binds guanine nucleotides (GTP with an association constant of 1 $\mu$M, GDP with a 5–10-fold lower affinity); (3) SelB binds selenocysteinyl-tRNA but is unable to bind this tRNA when it is charged with L-serine (Forchhammer et al. 1989).

The proposal that SelB is a specific translation factor required for the insertion of selenocysteine at the ribosome was supported further by the finding that EF-Tu has a very low affinity for selenocysteinyl-tRNA (Förster et al. 1990). EF-Tu and SelB, therefore, have a mutually exclusive function in delivering the 20 standard aminoacyl-tRNA's and selenocysteinyl-tRNA to the ribosome.

## 3 Conclusions and Perspectives

The pathway and the mechanism by which selenocysteine is synthesized and incorporated into polypeptides are unique and deviate in the following aspects from views accepted hitherto as being paradigms in biochemistry or molecular biology: (1) the results showed that the genetic code can accommodate amino acids in addition to the 20 standard ones; (2) the specificity of codon recognition is not just dependent on codon/anticodon interaction and on the immediate context of the particular codon but can be influenced by as much as 40 bases downstream on the mRNA, possibly via formation of a secondary structure. (3) There is a specific system which delivers the aminoacyl tRNA to the ribosome which is alternate to that of EF-Tu. In contrast to EF-Tu, this delivery protein, SelB, recognizes the aminoacyl-residue of the tRNA. Several of the unique structural features of this tRNA may have a function either in precluding its interaction with EF-Tu, and/or in assuring its preferential interaction with SelB. Elucidation of these features will not only aid in providing answers to some of the open questions concerning the mechanisms of polypeptide chain elongation and termination, but could also provide a means for insertion of selenocysteine in other proteins which normally do not contain this amino acid.

A question of particular interest is how this system may have arisen during evolution. The fact that eukaryotes seem to insert selenocysteine into glutathione peroxidase via a similar mechanism (Chambers et al. 1986; Lee et al. 1989) can be taken as circumstantial evidence, indicating that it is very old, having developed posssibly before eubacteria and eukaryotes separated. A possible (but not the only) interpretation is that the selenocysteine incorporation system is an evolutionary relic conserved from times when the protein synthesis machinery had not yet been streamlined to the extent which is evident today.

*Acknowledgment.* We are grateful to T.C. Stadtman, D. Söll, M. Sprinzl and M.-A. Mandrand-Berthelot for exchanging ideas and for fruitful cooperation. Work from the authors' laboratory has been supported by grants from the BMFT (Genzentrum München) and the Fonds der chemischen Industrie.

## References


Akasaka, M., Mizoguch, J., Yoshimura, S & Watanabe, K. (1989) Nucleotide sequence of cDNA for rabbit glutathione peroxidase. Nucl. Acids Res. 17:2136

Barrett, E.L., Jackson, E.C., Fukumoto, H.T. & Chang, G.W. (1979) Formate dehydrogenase mutants of *Salmonella typhimurium*: a new medium for isolation and new mutant classes. Mol. Gen. Genet. 177:95-101

Böck, A. & Stadtman, T.C. (1988) Selenocysteine, a highly specific component of certain enzymes, is incorporated by a UGA-directed co-translational mechanism. Biofactors 1:245-250

Brown, T.A. & Shrift, A. (1982) Selenium toxicity and tolerance in higher plants. Biol. Rev. 57:59-84

Chambers, I., Frampton, J., Goldfarb, P., Affara, N., McBain, W. & Harrison, P.R. (1986) The structure of the mouse glutathione peroxidase gene: the selenocysteine in the active site is encoded by the termination codon, TGA. EMBO J. 5:1221-1227

Cone, J.E., Martin del Rio, R., Davis, J.N. & Stadtman, T.C. (1976) Chemical characterization of the selenoprotein component of clostridial glycine reductase: identification of selenocysteine as the organoselenium moiety. Proc. Natl. Acad. Sci. USA 73:2659-2663

Doran, J.W. (1982) Microorganisms and the biological cycling of selenium. Adv. Microb. Ecol. 6:1-32

Dürre, P. & Andreesen, J.R. (1986) Die biologische Bedeutung von Selen. Biol. unserer Zeit 16:12-23

Enoch, H.G. & Lester, R.L. (1975) The purification and properties of formate dehydrogenase and nitrate reductase from *Escherichia coli*. J. Biol. Chem. 250:6693-6705

Epp, O., Ladenstein, R. & Wendel, A. (1983) The refined structure of the selenoenzyme glutathione peroxidase at 0.2 nm resolution. Eur. J. Biochem. 133:51-69

Förster, Ch., Ott, G., Forchhammer, K. & Sprinzl, M. (1990) Interaction of a selenocysteine-incorporating tRNA with elongation factor Tu from *E. coli*. Nucl. Acids Res. 18:487-491

Forchhammer, K., Leinfelder, W. & Böck, A. (1989) Identification of a novel translation factor necessary for the incorporation of selenocysteine into protein. Nature (Lond) 342:453-456

Forstrom, J.W., Zakowski, J.J. & Tappel, A.L. (1978) Identification of the catalytic site of rat liver glutathione peroxidase as selenocysteine. Biochemistry 17:2639-2644

Günzler, W.A., Steffens, G.T., Grossmann, A., Kim, S.-M.A., Ötting, F., Wendel, A. & Flohé, L. (1984) The amino-acid sequence of bovine glutathione peroxidase. Hoppe Seyler's Z. Physiol. Chem. 365:195-212

Hudman, J.F. & Glenn, A.R. (1984) Selenite uptake and incorporation by *Selenomonas ruminantium*. Arch. Microbiol. 140:252-256

Jones, J.B. & Stadtman, T.C. (1981) Selenium-dependent and selenium-independent formate dehydrogenases of *Methanococcus vannielii*. J. Biol. Chem. 256:656-663

Lee, B.J., Worland, P.J., Davis, J.N., Stadtman, T.C. & Hatfield, D.L. (1989) Identification of a selenocysteyl-$tRNA^{Ser}$ in mammalian cells that recognizes the nonsense codon, UGA. J. Biol. Chem. 264:9724-9727

Leinfelder, W., Forchhammer, K., Zinoni, F., Sawers, G., Mandrand-Berthelot, M.-A. & Böck, A. (1988a) *Escherichia coli* genes whose products are involved in selenium metabolism. J. Bacteriol. 170:540-546

Leinfelder, W., Zehelein, E., Mandrand-Berthelot, M.-A. & Böck, A. (1988b) Gene for a novel tRNA species that accepts L-serine and cotranslationally inserts selenocysteine. Nature (Lond) 331:723-725

Leinfelder, W., Stadtman, T.C. & Böck, A. (1989) Occurrence in vivo of selenocysteyl-$tRNA^{Ser}_{UCA}$ in *Escherichia coli*. J. Biol. Chem. 264:9720-9723

Leinfelder, W., Forchhammer, K., Veprek, B., Zehelein, E. & Böck, A. (1990) In vitro synthesis of selenocysteinyl-$tRNA_{UCA}$ from seryl-$tRNA_{UCA}$: Involvement and characterization of the *selD* gene product. Proc. Natl. Acad. Sci. USA 87:543-547

Lindblow-Kull, C., Kull, F.J. & Shrift, A. (1985) Single transporter for sulfate, selenate and selenite in *Escherichia coli* K12. J. Bacteriol. 163:1267-1269

Mandrand-Berthelot, M.-A., Wee, M.Y.K. & Haddock, B.A. (1978) An improved method for identification and characterization of mutants of *E. coli* deficient in formate dehydrogenase activity. FEMS Microbiol. Lett. 4:37-40

Menon, N.K., Peck, H.D., LeGall, J. & Przybyla, A.E. (1988) Cloning and sequencing of the gene encoding the large and small subunits of the periplasmic (NiFeSe) hydrogenase of *Desulfovibrio baculatus*. J. Bacteriol. 170:4429

Mizutani, T., Maruyama, N., Hitaka, T. & Sukenaga, Y. (1989) The detection of natural opal suppressor seryl-tRNA in *Escherichia coli* by the dot blot hydridisation and its phosphorylation by a tRNA kinase. FEBS Lett. 247:345–348

Mullenbach, G.T., Tabrizi, A., Irvine, B.D., Bell, G.I., Tainer, J.A. & Hallewell, R.A. (1988) Selenocysteine's mechanism of incorporation and evolution revealed in cDNAs of three glutathione peroxidases. Protein Engineering 2:239–246

Reddy, A.P., Hsu, B., Reddy, P.S., Li, N.-Q., Thyagaraju, U., Reddy, C.C., Tam, M.F. & Tu, C.-P.D. (1988) Expression of glutathione peroxidase I gene in selenium-deficient rats. Nucl. Acids Res. 16:5557–5568

Schön, A., Böck, A., Ott, G., Sprinzl, M. & Söll, D. (1989) The selenocysteine-inserting opal suppressor serine tRNA from *E. coli* is highly unusual in structure and modification. Nucl. Acids Res. 17:7159–7165

Stadtman, T.C. (1974) Selenium biochemistry. Science 183:915–922

Stadtman, T.C., Davis, J.N., Zehelein, E. & Böck, A. (1989) Biochemical and genetic analysis of *Salmonella typhimurium* and *Escherichia coli* mutants defective in specific incorporation of selenium into formate dehydrogenase and tRNAs. Biofactors 2:35–44

Voordouw, G., Menon, N.K., LeGall, J., Choi, E.-S., Peck, H.D. & Przybyla, A.E. (1989) Analysis and comparison of nucleotide sequences encoding the genes for [NiFe] and [NiFeSe] hydrogenases from *Desulfovibrio gigas* and *Desulfovibrio baculatus*. J. Bacteriol. 171:2894–2899

Wittwer, A.J. & Stadtman, T.C. (1986) Biosynthesis of 5-methylaminomethyl-2-selenouridine, a naturally occurring nucleoside in *E. coli* tRNA. Arch. Biochem. Biophys. 248:540–550

Yamamoto, I., Saiki, T., Liu, S.-M. & Ljungdahl, L.-V. (1983) Purification and properties of NADP-dependent formate dehydrogenase from *Clostridium thermoaceticum*, a tungsten-selenium-iron protein. J. Biol. Chem. 258:1826–1832

Yamazaki, S. (1982) A selenium-containing hydrogenase from *Methanococcus vannielii*. Identification of the selenium moiety as a selenocysteine residue. J. Biol. Chem. 257:7926–7929

Zinoni, F., Birkmann, A., Stadtman, T.C. & Böck, A. (1986) Nucleotide sequence and expression of the selenocysteine-containing polypeptide of formate dehydrogenase (formate-hydrogen-lyase-linked) from *Escherichia coli*. Proc. Natl. Acad. Sci. USA 83:4650–4654

Zinoni, F., Birkmann, A., Leinfelder, W. & Böck, A. (1987) Cotranslational insertion of selenocysteine into formate dehydrogenase from *E. coli* directed by a UGA codon. Proc. Natl. Acad. Sci. USA 84:3156–3160

Zinoni, F., Heider, J. & Böck, A. (1990) Features of the formate dehydrogenase mRNA necessary for decoding of the UGA codon as selenocysteine. Proc. Natl. Acad. Sci. USA 87:4660–4664

# Genetic Construction of Novel Metabolic Pathways: Degradation of Xenobiotica

K.N. TIMMIS[1]

## 1 Introduction

Photosynthetic microorganisms and plants are responsible for fixing carbon dioxide and thus for channeling inorganic carbon into the complex organic compounds of which living systems consist. The organic constituents of organisms are generally biodegraded, either to simpler organic compounds which serve as precursors for the synthesis of new complex molecules, or to inorganic carbon compounds such as $CO_2$, thereby completing the carbon cycle. Organic carbon may cycle through a number of different living organisms before finally being mineralized to $CO_2$. Soil and water microorganisms are important agents of biodecomposition and thus play a crucial role in recycling carbon and in maintaining the biosphere in equilibrium. Organic materials, however, may not necessarily be degraded but instead become converted to new organic compounds under the influence of geophysical and geochemical processes, and thereby exit from the carbon cycle for a period of time. The return to the carbon cycle of material sequestered in coal and oil deposits, etc. occurs through further geological processes and human activities that expose it to the biodegradative activities of microorganisms or that result in its combustion.

Whereas the number of different organic molecules resulting from biosynthetic processes is relatively small, and hence the spectrum of distinct pathways for the catabolism of such compounds limited, the variety of organic molecules in the biosphere which have been created by geological processes is enormous (Gibson and Subramanian 1984). It is not surprising therefore that the enterprising microbial world, which has evolved in such a way as to be able to profit maximally from the availability of a wide range of organic compounds (many of which are exotic or toxic to some degree), possesses an extremely varied arsenal of catabolic enzymes with which to attack potential growth substrates. This metabolic versatility reflects the exceptional genetic plasticity of microorganisms which facilitates the rapid evolution of new activities to take advantage of new nutritional sources. The genetic and metabolic versatility of microorganisms make them metabolic opportunists *par excellence.*

Superimposed on the rich variety of natural organic compounds present in the environment is the increasing number of novel industrial organic chemicals heavily substituted with chemical groups (halides, $SO_3$, $NO_2$, etc.) not found or not extensively present in organic compounds of biological origin. Whereas many of

[1]Division of Microbiology, GBF–National Research Centre for Biotechnology, W-3300 Braunschweig, FRG

41. Colloquium Mosbach 1990
The Molecular Basis of Bacterial Metabolism

these compounds can be processed by existing versatile catabolic pathways, or by readily occurring adaptations of existing pathways, others are so novel, toxic, or biologically (and chemically) resistant that they are not easily processed, and thus accumulate in the environment, sometimes causing severe pollution. One of the most important challenges of our time is to find and exploit existing biodegradative routes to eliminate pollutants from the environment or, when this is not possible, to develop experimentally new catabolic pathways that can be effective for this purpose.

There are a number of possible reasons why an organic compound may be biodegraded only slowly or not at all. One is that although the microbial world is genetically very fluid, the different genetic events that drive evolution (mutation, gene transfer, gene rearrangement and recombination, gene expression in new hosts) occur at specific frequencies that are typically of the order of $10^{-4}$–$10^{-8}$ per cell per generation, and that if multiple genetic events and several organisms are required to evolve a new activity, the time taken may be rather long. Currently available laboratory methods, however, allow the acceleration of these events provided we can predict which events are needed and can devise appropriate conditions to select for them: genes can be efficiently combined in precise configurations and subsequently transferred to appropriate host organisms at high frequencies.

If we are to construct microorganisms that exhibit desired catabolic activities, and expect them to express these activities in a predictable manner, we must first learn about the mechanisms of natural evolution and follow its rules. One feature of degradation pathways that has become apparent over the last few years is the relaxed substrate specificity of many catabolic enzymes and gene regulators (Harayama et al. 1986). This is presumably of crucial importance in the natural evolution of new catabolic activities. A goal of laboratory evolution strategies is therefore to build up a collection of genetic "modules" encoding key relaxed substrate specificity enzymes that can be rationally combined to produce desired metabolic routes for target substrates. In this brief overview, recent progress in the laboratory evolution of catabolic activities will be reviewed.

## 2 Strategies for in Vitro Evolution of Catabolic Pathways

In general, the evolution of new metabolic potential involves the acquisition of new or modified enzymatic activities (Campbell et al. 1973; Clarke 1978; Ramos et al. 1986, 1987a). However, since the synthesis of many enzymes is carefully regulated and occurs only in response to the appearance of specific induction signals (e.g., the appearance in the medium of the cognate substrate of the enzyme or pathway; Ramos et al. 1987b), the acquisition of new specificities of regulators of gene expression can also be important (Clarke 1978; Ramos et al. 1986, 1987a). De novo evolution of a protein with a radically new activity often involves a number of genetic events (mutations, recombinations, fusions, etc.), most of which cannot be predicted *a priori*. Thus, evolution of new phenotypes will generally involve the acquisition of new enzymatic or regulator specificities (for substrates and effectors, respectively) through mutational alteration of existing proteins or through recruitment of new proteins from different organisms. The ability of enzymes and regulators to undergo a relaxation of their specificities without loss of function, and the existence of enzymes exhibiting catalytic activities toward a broad range of structurally distinct substrate molecules are of critical importance to evolution.

Two general strategies have been developed for the laboratory evolution of catabolic pathways for recalcitrant compounds by means of genetic manipulation (Ramos and Timmis 1987). If the chemical exhibits substantial structural analogy to compounds that are readily degraded, the initial strategy of choice is to identify the steps of the known pathway that are nonpermissive for the chemical in question, and then to modify these such that they become permissive. This approach generally leads to an expansion of the substrate profile of the pathway. The expansion can be horizontal, in that more analogues of a single class of compounds are metabolized, as a result of the recruitment of isofunctional enzymes from other pathways (Lehrbach et al. 1984; see below) or of the mutational alteration of the substrate specificities of existing key enzymes (Clarke 1978; Ramos et al. 1986). Pathway expansion can also be vertical, in that the existing pathway is used as a base onto which are grafted additional enzymes that extend the pathway upward. Alternatively, if an existing pathway for related compounds is not known, new pathways can be conceived and appropriate component enzymes can be sought in bacteria found in natural habitats. Once the feasibility of a new route is established, the corresponding enzymes can be recruited into a single bacterium (or a consortium) by the cloning of their structural and regulatory genes and by the combination of such genes in appropriate host organisms (Rojo et al. 1987).

## 3 Rational Restructuring of Catabolic Pathways by Sequential Modification

*Pseudomonas putida* bacteria carrying TOL plasmid pWW0 are able to degrade and utilize, as sole source of carbon and energy, several alkylbenzenes including toluene, *m*- and *p*-xylenes, 3-ethyltoluene and 1,2,4-trimethylbenzene (Worsey and Williams 1975). Metabolism of these compounds is initiated by progressive oxidation of the methyl side chain substituent of carbon-1 (upper pathway), followed by oxidation of the aromatic carboxylic acid thereby formed to short chain carboxylic acids, pyruvate and aldehydes (*meta*-cleavage pathway; Fig. 1). Such bacteria can also grow on the corresponding alkylbenzoates but not, however, on either 4-ethyltoluene or 4-ethylbenzoate. Other bacteria, capable of metabolizing benzenes and benzoates, are readily isolated from soil but these also cannot generally degrade the 4-ethyl-substituted analogues. In order to identify the metabolic impediments to degradation of 4-ethylbenzene and to determine whether these may be overcome by appropriate genetic modification, we have carried out a substrate/effector study of the enzymes and regulators of the TOL catabolic pathway. In order to exploit more than once the growth of modified bacteria on 4-ethyl-substituted compounds as part of the strategy to select desired genetic changes, we initially focused on the *meta*-cleavage pathway before turning to the upper pathway.

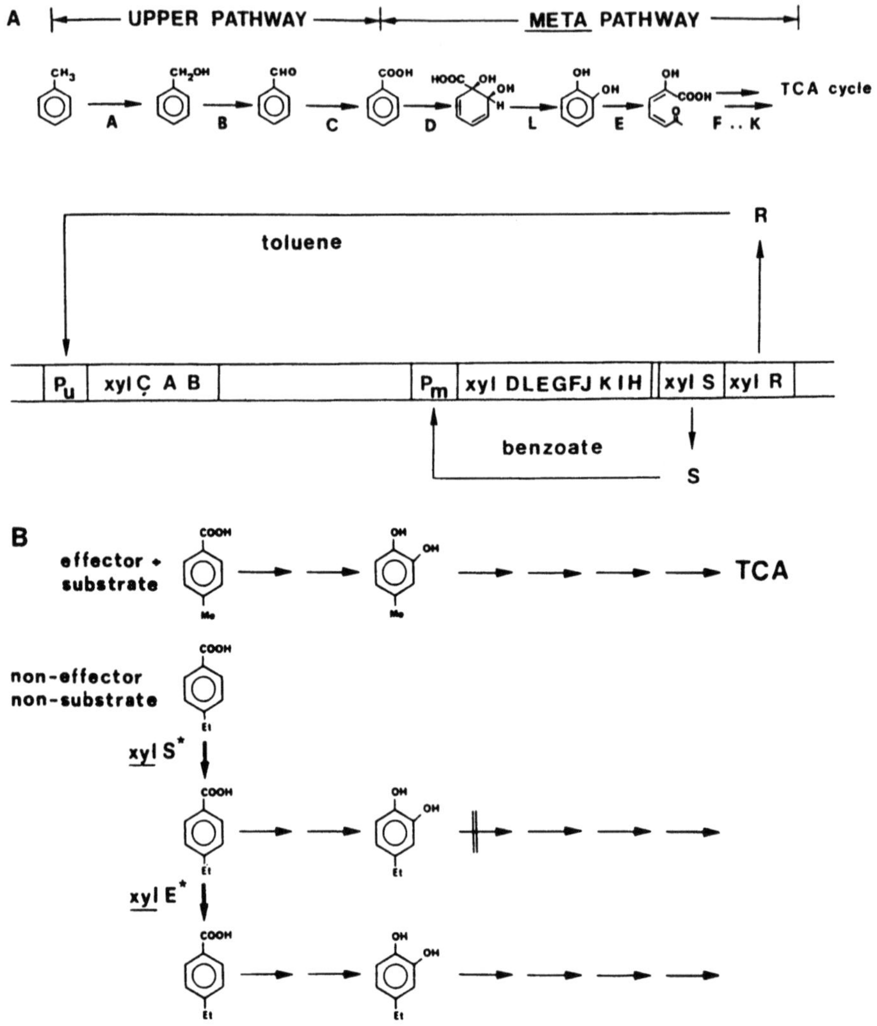


**Fig. 1A,B.** The TOL plasmid-encoded pathway for the degradation of alkylbenzenes and the experimental broadening of its substrate range. **A** The TOL catabolic pathway is shown for the substrate toluene, but 3- and 4-methyltoluene (*m*- and *p*-cresol, respectively), 1,2,4-trimethylbenzene, and 3-ethyltoluene are degraded in a similar fashion. The organization into two operons of the genes encoding the catabolic enzymes is shown, as are the regulatory circuits controlling transcription of these operons. The XylS positive regulator stimulates transcription of the *meta* operon when activated by benzoate or alkylbenzoates (i.e., the substrates of the *meta* operon), whereas the XylR positive regulator stimulates transcription of the upper operon when activated by toluene or another upper operon substrate. **B** 4-Methylbenzoate is both an effector molecule that activates the XylS protein, as well as a substrate for the *meta* pathway. 4-Ethylbenzoate does not serve as an effector for the XylS protein regulator of the catabolic operon promoter Pm (Table 1), and thus fails to induce synthesis of the catabolic enzymes. Isolation of a mutant that produces a XylS protein analogue, which is activated by 4-ethylbenzoate, results in the synthesis of all catabolic enzymes in response to the presence of 4-ethylbenzoate in the bacterial culture medium. However, this compound is only metabolized as far as 4-ethylcatechol, because the latter inactivates the ring cleavage enzyme C23O. The isolation of a mutant that produces a C23O analogue resistant to inactivation by 4-ethylcatechol eliminates the final metabolic block and permits complete degradation of 4-ethylbenzoate through the *meta* pathway

Expression of the TOL plasmid genes is regulated at the transcriptional level. The upper pathway operon is induced by toluene/xylenes and their alcohol derivatives and this induction is mediated by the *xylR* gene product positive regulator (Fig. 1). Expression of the *meta*-cleavage operon is induced by benzoate/toluates, and this induction is mediated by the *xylS* gene product positive regulator. The promoters of the two operons and their accompanying regulatory determinants have been localized and sequenced, and the operon transcriptional start sites have been identified (Ramos et al. 1987b).

In order to examine the inducer specificity of the positive regulator of expression of the *meta*-cleavage operon, the promoter of this operon *Pm* was fused in a cloning vector to *lacZ*, which encodes β-galactosidase, a readily assayed test enzyme used extensively in gene expression studies (Ramos et al. 1986). *Escherichia coli* K-12 carrying this hybrid plasmid, plus another containing *xylS*, the gene of the positive regulator of transcription from *Pm*, synthesizes β-galactosidase when benzoate is present in the growth medium but not when a noninducer such as salicylate is present. Using this system, a large number of benzoate analogues were tested for their ability to provoke synthesis of β-galactosidase through activation of the XylS regulator. The structural features of benzoate analogues critical for their interaction with the regulator were thereby analyzed (Table 1). From this study, it became clear that whereas a number of benzoate analogues

**Table 1.** Activation of XylS and a XylS mutant protein by benzoate analogues[a]

| | Induction ratio (relative increase in β-galactosidase) | |
|---|---|---|
| Benzoate analogue | XylS | XylS4 |
| None | 1 | 1 |
| 2MB | 18 | 37 |
| 3MB | 17 | 14 |
| 4MB | 4 | 10 |
| 2,3MB | 10 | 12 |
| 3,4MB | 5 | 8 |
| 2,4MB | 1 | 5 |
| 2,5MB | 1 | 1 |
| 3,5MB | 1 | 1 |
| 4EB | 1 | 8 |

[a] *Escherichia coli* K-12 bacteria containing two plasmids, one carrying a *Pm*::*lacZ* fusion and the other carrying either *xylS* or the mutant *xylS* allele *xylS4* were cultured in L-broth containing or lacking the benzoate analogue indicated in the left-hand column. β-Galactosidase levels were subsequently measured, and the ratio plus analogue/minus analogue (induction ratio) was determined. A value of 1 indicates that the analogue did not induce synthesis of β-galactosidase, and thus does not serve as an effector of the XylS or mutant XylS proteins, whereas values greater than 1 indicate that the analogue served as an inducer. Abbreviations: B, benzoate; M, methyl; E, ethyl.

activate the XylS protein and induce synthesis of the *meta*-pathway enzymes, 4-ethylbenzoate does not (Ramos et al. 1986).

In order to select *xylS* mutants that produce regulators having broader inducer specificities, a hybrid plasmid was constructed in which the *Pm* promoter was fused to a gene specifying resistance to tetracycline (Fig. 2). This gene is expressed, and thus host bacteria are resistant to tetracycline, *only* when the XylS protein is activated by benzoate or one of its analogues. Bacteria spread on a nutrient agar plate containing tetracycline and 4-ethylbenzoate (or some other ordinarily noninducing benzoate analogue) are killed because the XylS protein is not activated. The plating of a large number of such bacteria and the inclusion of a strong mutagen such as ethylmethane sulfonate in the plate result in the isolation of a few tetracycline-resistant colonies (Fig. 2). Some of these colonies produce mutant XylS proteins that are activated by new effector molecules such as 4-ethylbenzoate (Ramos et al. 1986, 1987a, 1990).

One mutant *xylS* gene, *xylS4* (present on pERD4), was transferred to *P. putida* bacteria carrying the TOL plasmid (Fig. 2). This derivative also failed to grow on 4-ethylbenzoate but did degrade the compound to 4-ethylcatechol (Fig. 1). Thus, the catabolic enzymes are synthesized in this derivative, and 4-ethylbenzoate is transformed to 4-ethylcatechol, but the *meta*-cleavage enzyme does not permit further metabolism (Ramos et al. 1987a). Characterization of the *meta*-cleavage enzyme, catechol 2,3-dioxygenase (C23O), revealed that 4-ethylcatechol is in fact a suicide substrate which causes irreversible inactivation of the enzyme.

It was reasoned that if a single amino acid change in C23O could render it resistant to inactivation by 4-ethylcatechol, and that if this were the only further change needed to permit the complete degradation of 4-ethylbenzoate via the TOL plasmid-encoded pathway, then the appropriate C23O mutants might be obtained by selecting directly for growth of bacteria on 4-ethylbenzoate. A large number of *P. putida* bacteria carrying the TOL plasmid and the *xylS4* gene were plated on minimal medium containing ethylmethane sulfonate and 4-ethylbenzoate as the sole source of carbon and energy (Fig. 2). Several colonies grew under these conditions and all were subsequently shown to be able to use 4-ethylbenzoate as a source of carbon and energy, and to produce altered C23O enzymes that exhibited increased resistance to inactivation by 4-ethylcatechol (Ramos et al. 1987a).

Although *P. putida* bacteria carrying a derivative TOL plasmid containing the *xylS4* mutation and the *xylE* mutation conferring 4-ethylcatechol resistance upon C23O are able to grow on 4-ethylbenzoate as sole carbon source, they are not able to grow on 4-ethyltoluene. An analogous approach was therefore taken to analyze

**Fig. 2A,B.** Sequential mutational elimination of metabolic barriers to the degradation of 4-ethylbenzoate through the TOL plasmid-specified *meta* pathway. **A** Selection of mutant XylS proteins activated by new effector molecules such as 4-ethylbenzoate. *Escherichia coli* K-12 (pNM185; pJLR200) bacteria were spread on nutrient agar plates containing ampicillin, kanamycin, tetracycline, 4-ethylbenzoate, and the mutagen ethylmethane sulfonate (EMS). After incubation of the plates two classes of clones appeared: one class that grew on tetracycline plates lacking 4-ethylbenzoate (most of these contained a mutation in the *Pm* promoter, so that the *tet* gene was expressed constitutively), and another class that grew only on tetracycline plates containing 4-ethylbenzoate. Mutants of the latter type

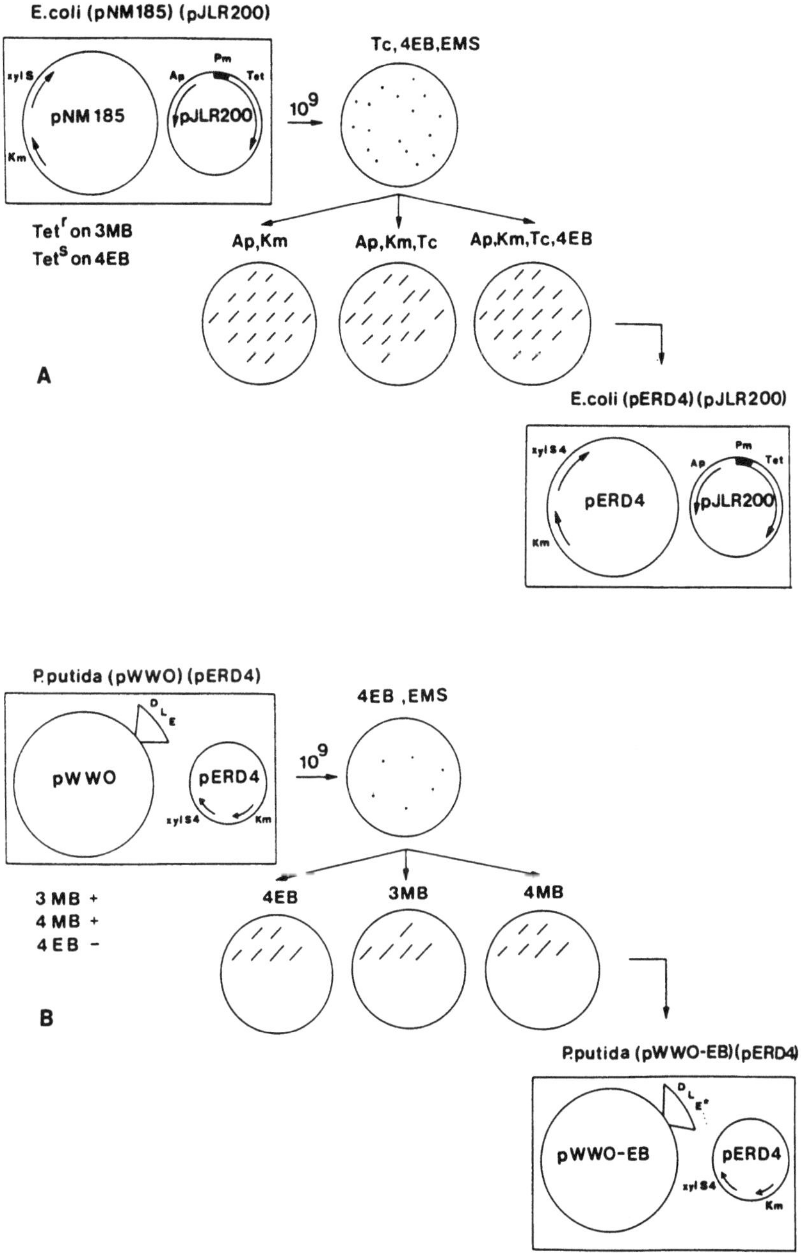


contained plasmids such as pERD4 that carry mutant *xylS* genes (e.g., *xylS4*) whose products are activated by 4-ethylbenzoate (Table 1). **B** Isolation of mutants producing C23O analogues resistant to inactivation by 4-ethylcatechol. *Pseudomonas putida* (pWW0; pERD4) bacteria were spread on minimal medium containing 4-ethylbenzoate as sole source of carbon and energy and the mutagen EMS. After incubation of the plates, six clones appeared, five of which grew on 4-ethylbenzoate and 3- and 4-methylbenzoate, while the other grew only on 4-ethyl- and 4-methylbenzoate. All clones contained mutant *xylE* genes (*xylE**) that encode C23O enzyme analogues exhibiting resistance to inactivation by 4-ethylcatechol

the effector and substrate specificities of the XylR regulator and catabolic enzymes, respectively, of the upper pathway.

A fusion between *lacZ* and the upper pathway operon promoter *Pu* permitted the analysis of the effector specificity of the XylR protein. This study revealed that the effector specificity of this protein is extremely relaxed and that XylR is activated by a wide range of alkyl- and chlorotoluenes, including 4-ethyltoluene, and by methyl- and chlorobenzylalcohols (Abril et al. 1989). Thus, failure to metabolize 4-ethyltoluene is not due to a lack of synthesis of the catabolic enzymes.

Analysis of the substrate specificities of the three upper pathway enzymes, toluene oxidase, benzylalcohol dehydrogenase, and benzaldehyde dehydrogenase revealed that although the latter two had rather relaxed substrate specificities, and transformed the corresponding 4-ethyl-substituted intermediates, toluene oxidase had little to no activity on chlorotoluenes or 4-ethyltoluene. It seemed therefore that the remaining barrier to 4-ethyltoluene degradation in such bacteria resided in the narrow substrate specificity of a single enzyme, the first in the pathway. Again reasoning that a single mutation might eliminate this barrier and enable bacteria to metabolize 4-ethyltoluene, we plated a large number of bacteria on minimal medium containing the mutagen EMS and having 4-ethyltoluene as sole source of carbon and energy. Mutants growing on this medium arose at a frequency of about $10^{-8}$. Such mutants transferred their new phenotype to other bacteria along with the TOL plasmid and produced an altered toluene oxidase that attacks 4-ethyltoluene (Abril et al. 1989).

The TOL plasmid-encoded pathway for the degradation of alkylbenzenes had thereby been rationally restructured to enable it to process the rather recalcitrant analogue 4-ethyltoluene by sequential mutational modification of three elements, resulting in relaxation of the effector specificity of the XylS protein, an increase in the resistance of catechol 2,3-dioxygenase to inactivation by 4-ethylcatechol, and a broadened substrate specificity of the initial enzyme of the pathway, toluene oxidase.

## 4 Designing New Catabolic Pathways: Patchwork Assembly of Enzymes and Regulatory Systems

The redesigning of an existing pathway is a relatively straightforward experimental approach to evolve a pathway for the degradation of a recalcitrant compound. However, this approach will not always be sufficient, in which case it may be necessary to construct a new pathway by rationally combining in patchwork fashion a series of catalytic activities derived from different pathways and different organisms.

Individual substituted phenols and benzoates have a significant level of toxicity, but nevertheless can generally be degraded via catechol by microorganisms in soil and in wastewater treatment plants. Industrial wastes, however, frequently contain mixtures of chloro- and methyl-substituted phenols and benzoates. Such mixtures are not only difficult to degrade but also tend to destabilize phenol- and benzoate-degrading microbial communities. The problem lies in the existence of both *ortho*- and *meta*-cleavage routes for degradation of catechols.

Catechol and chlorocatechols are generally subjected to *ortho*-fission, whereas methylcatechols suffer *meta*-fission (Fig. 3). Although both pathways may exist in individual microorganisms, only one is usually functional at any given moment, according to the substrate that is available. However, when both chloro- and methylcatechols are formed from mixtures of chloro- and methylaromatics, both types of pathways are functional, often in the same organism, and the catechols will be subjected to both types of cleavage. Whereas *ortho*-cleavage of chlorocatechols leads to their productive metabolism, *ortho*-cleavage of methylcatechols leads to the formation of dead-end products (Fig. 3). Similarly, whereas the *meta*-cleavage of methylcatechols leads to their productive metabolism, the *meta*-cleavage of chlorocatechols leads to the formation of either dead-end products or reactive products that inactivate C23O, the ring cleavage enzyme (Knackmuss 1983). The nonproductive misrouting of catechol cleavage products during simultaneous metabolism of chloro- and methyl-substituted aromatics constitutes a sort of biochemical anarchy and eventually perturbs the productive metabolism of aromatics by the cell or the community to such an extent that cell death or disruption of the community may occur (Knackmuss 1983; Rojo et al. 1987; Fig. 3).

One potential solution to this problem is the construction of catabolic routes for chloro- and methylaromatics that employ only one type of catechol ring fission mode. *Pseudomonas* sp. B13 possesses only *ortho*-cleavage routes for catechols; it can grow on 3-chlorobenzoate and acquires the ability to grow on 4-chlorobenzoate when it recruits the TOL plasmid-encoded relaxed substrate specificity toluate dioxygenase, through transfer of the *xylDLS* genes of the TOL plasmid (Reinecke and Knackmuss 1979; Lehrbach et al. 1984). In order to create a stable B13 derivative, which is able to degrade 4-chlorobenzoate, the cloned TOL genes were inserted into transposon Tn*5*, and the hybrid element was subsequently transposed into the B13 chromosome (Rojo et al. 1987). The derivative thereby obtained, FR1 (Fig. 4), grew on 3- and 4-chlorobenzoate but not on 3- or 4-methylbenzoate. The latter two compounds were, however, cometabolized via *ortho*-cleavage to the dead-end products 2- and 4-methyl-2-enelactone, respectively. In order to effect mineralization of the dead-end products of *ortho*-cleavage of 3- and 4-methylbenzoate, it was necessary to identify and recruit additional enzymes. The methyl-2-enelactones that accumulated during metabolism of methylbenzoates were therefore isolated and used to select organisms able to use such compounds for growth. This resulted in the isolation of an *Alcaligenes* sp. capable of growing on 2- and 4-methyl-2-enelactone. Growth on 4-methyl-2-enelactone involves isomerization of 4-methyl-2-enelactone to 3-methyl-2-enelactone. Since *Pseudomonas* sp. B13 is able to grow on 3-methyl-2-enelactone, it was reasoned that recruitment of the isomerase function of the *Alcaligenes* sp. into FR1 would allow it to grow on 4-methylbenzoate. An *Alcaligenes* gene bank was therefore prepared in a wide host range, mobilizable cosmid vector and mass-transferred into FR1 by conjugation. Transconjugants that were able to grow on 4-methylbenzoate were readily isolated; the hybrid cosmid present in one of these was designated pFRC20P. Cell-free extracts of these transconjugants exhibited enzymatic activities that converted purified 4-methyl-2-enelactone to 3-methyl-2-enelactone. High levels of activity were measured both in 4-

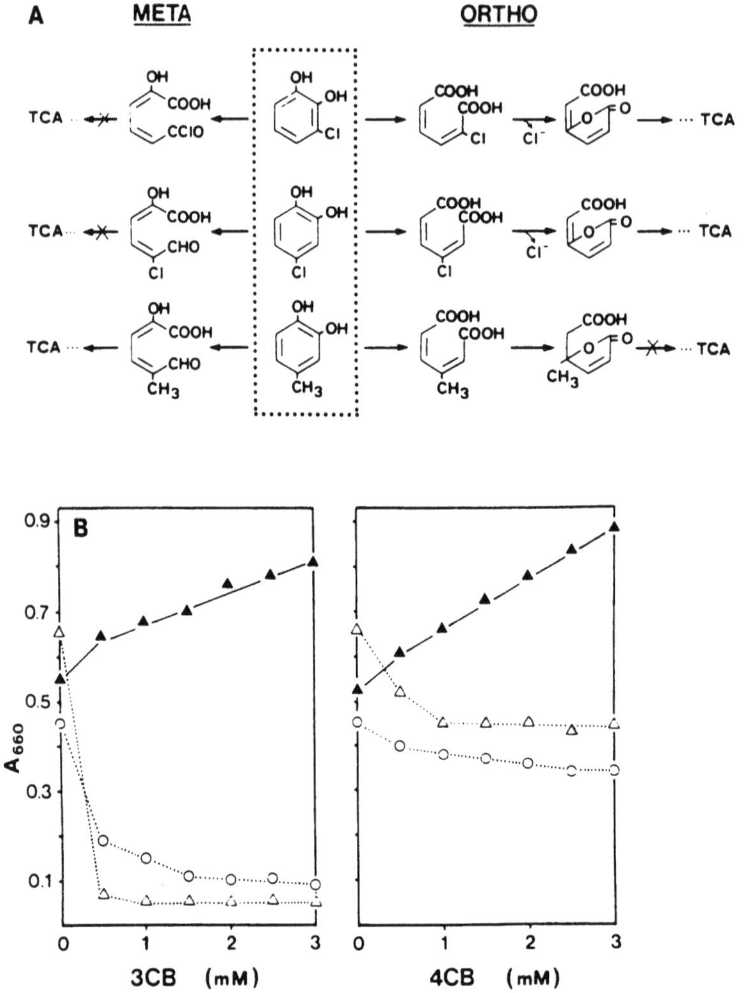


**Fig. 3A,B.** Unproductive misrouting of substituted catechols by *ortho*- and *meta*-cleavage enzymes. **A** Degradation of chloro- and methyl-substituted aromatics often involves as a first step their conversion to corresponding catechol derivatives. Chlorocatechols are degraded through *ortho*-cleavage pathways, but may be misrouted into a *meta*-cleavage pathway in bacteria containing a *meta*-cleavage enzyme. As a result, they are channeled to dead-end metabolites which in some cases (such as 3-chlorocatechol) are highly reactive and inactivate the *meta*-cleavage enzyme. On the other hand, methylcatechols are normally degraded through *meta*-cleavage routes, although they can be transformed by *ortho*-cleavage pathways, eventually forming methyllactone derivates as dead-end metabolites. **B** Effect of misrouting substituted catechols. Bacteria were grown in minimal medium containing 4-methylbenzoate (5 mM) as sole source of carbon and energy. When the cultures reached an absorbance of $A_{660}$ = 0.5, they were diluted tenfold with fresh medium containing both 4-methylbenzoate and either 3-chlorobenzoate (*left*) or 4-chlorobenzoate (*right*) at the indicated concentrations, and incubation was continued for a further 20 h. The final $A_{660}$ was measured and plotted against the concentration of 3- and 4-chlorobenzoate initially present in the fresh medium. The strains used were: *Pseudomonas putida* KT2440 (TOL) (○; possesses a *meta* pathway for degradation of methylbenzoates); *Pseudomonas* sp. B13(TOL) (△; possesses TOL plasmid *meta* pathway and an *ortho* pathway for 3-chlorobenzoate; B13 derivative FR1 (pFRC20P) (▲; lacks a *meta* pathway, but possesses *ortho* pathways for degradation of both chloro- and methylbenzoates)

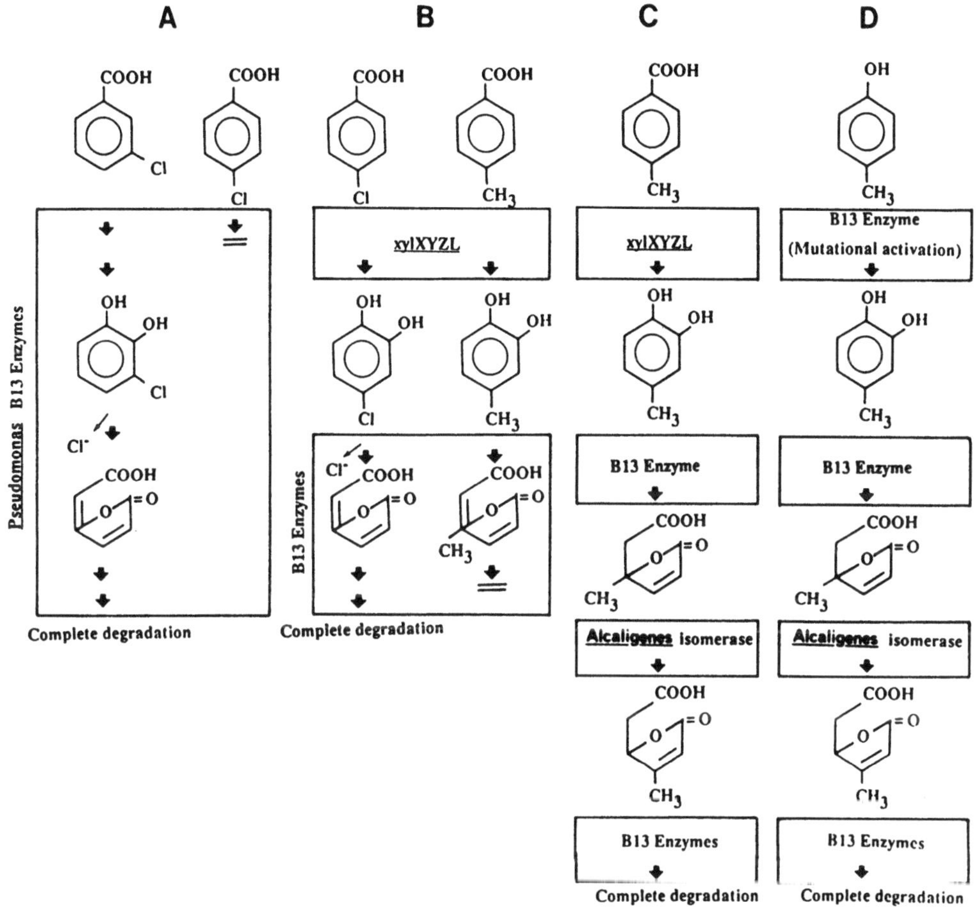


**Fig. 4A-D.** Constructed pathway for the simultaneous degradation of chloro- and methylaromatics. The route is based on the modified *ortho* pathway for 3-chlorobenzoate of *Pseudomonas* sp. B13 (**A**). Introduction into B13 of the TOL plasmid genes coding for toluate 1,2-dioxygenase (*xylXYZ*, which also has the single allele designation *xylD*) and dihydroxycyclohexadiene carboxylate dehydrogenase (*xylL*), together with that of the positive regulator of the xylDL operon (*xylS*), expands the degradation range to include 4-chlorobenzoate and permits transformation of methylbenzoates to methyl-2-enelactones, which are accumulated as dead-end metabolites (**B**). Recruitment of a 4-methyl-2-enelactone isomerase from *Alcaligenes eutrophus* allows the transformation of 4-methyl-2-enelactone (the intermediate of 4-methylbenzoate catabolism) to 3-methyl-2-enelactone (*bottom right*), which is completely degraded by other enzymes of B13 (**C**). Mutational modification of the expression of the phenol hydroxylase of B13 further extends the degradation capacities to chloro- and methylphenols (**D**). The final route is thereby composed of five different pathway segments derived from three different organisms

methylbenzoate-grown cells and in acetate-grown cells, although highest levels were obtained in 4-methylbenzoate-grown cells. This indicates that expression of the isomerase is regulated in these constructions, but that the basal level of synthesis is high.

*Pseudomonas* sp. B13 is able to grow on phenol and, after adaptation, on 4-chlorophenol as sole sources of carbon and energy; catabolism is via an *ortho*-cleavage route, with catechol or 4-chlorocatechol as intermediates (Knackmuss and Hellwig 1978). The phenol hydroxylase involved also transforms 3- and 4-methylphenols in cell-free extracts and produces from these substrates the corresponding methylcatechols. In principle, therefore, an appropriate derivative of B13 such as FR1(pFRC20P) that can mineralize 4-methylcatechol possesses all of the enzymes necessary to grow on 4-methylphenol as sole source of carbon and energy. However, although phenol is a growth substrate for this derivative, 4-methylphenol is not. Spontaneous mutants could nevertheless be selected that grow on 4-methylphenol (Fig. 4; frequency $10^{-7}$ to $10^{-8}$). Measurement of phenol hydroxylase levels in such mutant bacteria grown in succinate, phenol, or 4-methylphenol revealed little or no enzyme in succinate-grown cells and high activities in bacteria grown on phenol or 4-methylphenol. Synthesis of this enzyme is therefore specifically regulated.

Figure 3 shows that addition of a chloroaromatic such as 3-chlorobenzoate to bacteria [e.g. *P. putida* KT2440(TOL) or *Pseudomonas* sp. B13(TOL)] that are actively degrading a methylbenzoate through a *meta* ring-fission pathway resulted in inhibition of cell growth as a result of misrouting of 3-chlorocatechol into the *meta* pathway and irreversible inactivation of the key enzyme of the pathway, C23O, by the ring-fission product of 3-chlorocatechol, which is a highly reactive acylhalide. Addition of 4-chlorobenzoate to such cells had a less severe impact on growth because the dead-end product that forms as a result of misrouting of 4-chlorocatechol into the *meta* pathway is less reactive (Fig. 3). Addition of 3- or 4-chlorobenzoate to FR1(pFRC20P) bacteria growing on 4-methylbenzoate had no inhibitory effect and in fact promoted further growth of the cultures (Fig. 3). Moreover, the B13 derivative degraded simultaneously the chloro- and methylbenzoates in the given mixtures.

Thus, a novel *ortho*-cleavage route for the degradation of mixtures of 3- and 4-chloro- and 4-methylbenzoates has been constructed by the patchwork assembly of four pathway segments (Fig. 4), namely: (1) the TOL plasmid toluate dioxygenase and following *cis*-diol dehydrogenase, which transforms methylbenzoates to methylcatechols; (2) the B13 catechol-1,2-dioxygenase and muconate cycloisomerase, which convert methylcatechols to methyl-2-enelactones; (3) the *Alcaligenes eutrophus* 4-methyl-2-enelactone isomerase, which converts 4-methyl-2-enelactone to 3-methyl-2-enelactone; and (4) the B13 3-methyl-2-enelactone pathway, which completes the catabolic route. This pathway was further expanded through activation of the synthesis of a relaxed substrate specificity phenol hydroxylase (e). The newly evolved strain metabolizes chloro- and methyl-substituted phenols and benzoates exclusively via *ortho*-cleavage routes, tolerates shock loads of one type of substituted benzoate or phenol while utilizing the other as a carbon source, and degrades both types simultaneously when present together.

The constructed strain is a useful model to analyze the efficiency and stability of new pathways constructed in the laboratory since some of the determinants of the pathway have been integrated into the bacterial chromosome (those of toluate dioxygenase), some exist on the B13 plasmid pWR1 (chlorocatechol pathway), and some on a recombinant plasmid (methyllactone isomerase). The strain was added to contained microcosms fed with a synthetic sewage mix with or without chloro- and methylbenzoate and its survival and activity monitored over 10-day periods (Dwyer et al. 1988). These experiments revealed that the added bacteria rapidly dropped in number from $10^8$/ml to about $10^5$/ml but persisted at this level for the duration of the experiment, and that they protected the microcosms from the deleterious effects of shock loads of chloro- and methylbenzoate mixtures. Stability studies have shown the recombinant plasmid containing the isomerase to be relatively unstable in the absence of continued selection pressure. We therefore subcloned the isomerase gene into a transposon and integrated the recombinant transposon into the B13 chromosome. This strain is highly stable for the 4-methylbenzoate degradation phenotype, more stable in fact than for the intrinsic but pWR1 plasmid-encoded 3-chlorobenzoate degradation property (S. Negoro et al. submitted).

## 5 Concluding Remarks

Current levels of environmental pollution urgently necessitate, among other measures, much greater exploitation of microbial degradative activities. Unfortunately, some chemicals possess structural elements or substituents that confer upon the molecule a high degree of resistance to enzymatic attack, or are present in mixtures that are incompatible with the effective degradation of the toxic component. Soil and water microorganisms have the capacity to evolve enzymes capable of attacking most classes of organic chemicals, but the evolution of new catabolic pathways proceeds very slowly where multiple genetic changes are required, and where the selection pressures may only be effective in selecting the last genetic change to occur.

A major advantage of experimental evolution of pathways is that laboratory selection conditions (e.g., antibiotic treatment when selecting acquisition of a hybrid plasmid containing the gene of an enzyme to be recruited or a mutant regulatory protein; see above) may be totally unrelated to the ultimate phenotypic change desired (e.g., catabolism of an aromatic compound). Effective selection procedures can thus be *custom designed* for each of the individual genetic changes required. Successive changes can either be effected in the organisms to be evolved, or be effected in different organisms and subsequently combined sequentially or simultaneously in the organism selected to carry out the biodegradative activity. In this way, the evolutionary process can be substantially accelerated. The cloning of genes of proteins having useful characteristics, such as relaxed substrate specificities, both for the purpose of genetic and functional analysis and for their transfer into organisms to be evolved, generates "evolutionary modules" for further experiments. Growth in the number of different modules available will increasingly facilitate the design and experimental evolution of new and more complex pathways.

In the experiments described above, existing pathways were restructured by mutational alteration of protein specificities so that the modified proteins recognized new substrates/inducers, and novel metabolic routes were created by the patchwork assembly of enzymes and regulators from different pathways and from different organisms. These experiments confirm that experimental evolution of metabolic pathways is feasible and holds considerable potential for accelerating the evolution of microbes that will be able to degrade particularly recalcitrant and toxic compounds.

Although we have focused largely on the manipulation of genes of enzymes and regulators, it is evident that other targets for experimental intervention will become relevant in the near future. The recruitment of appropriate membrane transport systems for unusual growth or transformation substrates may, in some cases, be important, as will the isolation of mutants resistant to the toxicity of certain compound(s). In addition, however, there may be applications in which the efficacy of evolved activities can be considerably increased if the manipulated bacterium is able to undertake vectorial movement toward a nondiffusible or poorly diffusible target (e.g., plant root, hydrophobic chemical). Thus, experimental manipulation of bacterial taxis and motility should receive increasing attention during the next few years.

*Acknowledgments.* It is a pleasure to acknowledge the many fruitful and stimulating discussions with colleagues of the Department of Microbiology at the GBF and H.-J. Knackmuss and his group at the University of Stuttgart, and the excellent secretarial assistance of S. Ludwig and H. Brink.

## References

Abril, M.A., Michan, C., Timmis, K.N. & Ramos, J.L. (1989) Regulator and enzyme specificities of the TOL plasmid-encoded upper pathway for degradation of aromatic hybrocarbons and exponsion of the substrate range of the pathway. J. Bacteriol. 171:6782–6790

Campbell, J.H., Lengyel, J.A. & Langridge, J. (1973) Evolution of a second gene for $\beta$-galactosidase in *Escherichia coli*. Proc. Natl. Acad. Sci., USA 70:1841–1845

Clarke, P.H. (1978) Experiments in microbial evolution. In: Ornston, L.N. & Sokatch J.R. (eds) The Bacteria, vol. 4. Academic Press, New York, pp. 137–218

Dwyer, D.F., Rojo, F. & Timmis, K.N. (1988) Fate and behaviour in an activated sludge microcosm of a genetically-engineered microorganism designed to degrade substituted aromatic compounds. In: Sussmann, M., Collins, C.H., Skinner, F.A. & Stewart, D.E. (eds) The Release of Genetically-Engineered Macroorganisms. Academic Press, London, pp. 77–88

Gibson, D.T. & Subramanian, V. (1984) Microbial degradation of aromatic hydrocarbons. In: Gibson, D.T. (ed) Microbial degradation of organic compounds. Microbiology Series, vol. 13. Marcel Dekker, New York, pp. 181–252

Harayama, S., Ramos, J.L. & Timmis, K.N. (1986) Experimental evolution of plasmid specified functions. In: Levy, S.B. & Novick, R.P. (eds) Antibiotic resistance genes: ecology, transfer and expression. Banbury Report 24, Cold Spring Harbor Laboratory, Cold Spring Harbor, New York, pp. 389–402

Knackmuss, H.J. & Hellwig, M. (1978) Utilization and cooxidation of chlorinated phenols by *Pseudomonas* sp. strain B13. Arch. Microbiol. 117:1–7

Knackmuss, H.J. (1983) Xenobiotic degradation in industrial sewage: haloaromatics as target substrates. In: Phelps, C.F. & Clarke, P.H. (eds) Biotechnology, Biochemical Society Symposium No. 48, Biochemical Society, London, pp. 173-190

Lehrbach, P.R., Zeyer, J., Reineke, W., Knackmuss, H.J. & Timmis, K.N. (1984) Enzyme recruitment in vitro: use of cloned genes to extend the range of haloaromatics degraded by *Pseudomonas* sp. strain B13. J. Bacteriol. 158:1025-1032

Ramos, J.L. & Timmis, K.N. (1987) Experimental evolution of catabolic pathways of bacteria. Microbiol. Sci. 4:228-237

Ramos, J.L., Stolz, A., Reineke, W. & Timmis, K.N. (1986) Altered effector specificities in regulators of gene expression: TOL plasmid *xylS* mutants and their use to engineer expansion of the range of aromatics degraded by bacteria. Proc. Natl. Acad. Sci. USA 83:8467-8471

Ramos, J.L., Wasserfallen, A., Rose, K. & Timmis, K.N. (1987a) Redesigning metabolic routes: manipulation of TOL plasmid pathway for catabolism of alkybenzoates. Science 235:593-596

Ramos, J.L., Mermod, N. & Timmis, K.N. (1987b) Regulatory circuits controlling transcription of TOL plasmid operon encoding *meta*-cleavage pathway for the degradation of alkylbenzoates by *Pseudomonas*. Mol. Microbiol. 1:293-300

Ramos, J.L., Michan, C., Rojo, F., Dwyer, D. & Timmis, K.N. (1990) Signal-regulator interactions. Genetic analysis of the effector binding site of *xylS*, the benzoate activated positive regulator of *Pseudomonas* TOL plasmid *meta*-cleavage pathway operon. J. Mol. Biol. 211:373-382

Reineke, W. & Knackmuss, H.J. (1979) Construction of haloaromatics utilizing bacteria. Nature 277:385-386

Rojo, F., Pieper, D.H., Engesser, K.H., Knackmuss, H.J. & Timmis, K.N. (1987) Assemblage of *ortho* cleavage route for simultaneous degradation of chloro- and methylaromatics. Science 235:1395-1398

Worsey, M.J. & Williams, P.A. (1975) Metabolism of toluene and the xylenes by *Pseudomonas putida* (arvilla) mt-2: evidence for a new function of the TOL plasmid. J. Bacteriol. 127:7-13

# Elements of Quinone-Mediated Redox Catalysis in the Photosynthetic Reaction Center Protein

K. Warncke[1], M.R. Gunner[1], K.M. Giangiacomo[1], J.M. Keske[1], J.M. Bruce[2] and P.L. Dutton[1]

## 1 Introduction

Quinone cofactors play a central role in the photosynthetic and respiratory electron transport systems of eukaryotes and prokaryotes (Dutton 1986). Quinone present in the native bioenergetic membranes in large excess ($Q_{pool}$) over the charge-separating protein complexes serves as the carrier of electrons and protons between the complexes by cycling between the oxidized (Q) and two-electron reduced and protonated hydroquinol ($QH_2$) species (Hauska and Hurt 1982). The interface between Q and $QH_2$ in the $Q_{pool}$ and the redox proteins occurs at distinct binding sites (Trumpower 1982). Due to the sequential, single-electron nature of quinone reduction and oxidation, a common feature of the catalytic mechanism of these redox sites is the one-electron reduced, anionic semiquinone ($Q^-$) state.

Our efforts to elucidate the factors contributing to $Q/Q^-$-mediated redox catalyses has focused in recent years on the electrochemically and functionally distinct $Q_A$ and $Q_B$ sites of the reaction center protein (RC) from the bacterium *Rhodobacter sphaeroides*, which can be conveniently studied in the same system. The $Q_A$ site quinone is restricted by protein to one-electron redox behavior ($Q/Q^-$), whereas the $Q_B$ site conducts two sequential, single-electron reduction steps followed by direct binding of two protons ($Q/Q^-$, $Q^-/Q^{-2}$; $Q^{-2}/QH_2$) (Wraight 1982). This behavior is tailored to the role of the $Q_A$ quinone as terminal acceptor in the rapid, charge-separating sequence of primary intraprotein electron transfers, and $Q_B$ as the interface between the RC and the $Q_{pool}$.

The experimental approaches center on replacement of the native ubiquinone-10 cofactors (Okamura et al. 1975) with a variety of quinone and analog molecules in order to assess: (1) the bases of binding interaction and site specificity, from comparisons of the binding free energies ($\Delta G^0_B$) (from dissociation constants, $K_D$, determined in water and hexane solutions) (Gunner et al. 1985, 1990; Giangiacomo et al. 1990; Warncke et al. 1990), and (2) factors contributing to the selective solvation of bound Q and $Q^-$, from values of the electrochemical midpoint potentials of the couples [$E_m$(in situ)] (Woodbury et al. 1985; Giangiacomo et al. 1990). This approach is supported by in vitro partitioning (Braun et al. 1986; Keske et al. 1990) and electrochemical studies in various solvents (Prince et al. 1982; Warncke and Dutton 1989) to address features of quinone solvation as they relate to redox transformations in the catalytic sites. The ultimate aim of this work is to

[1] Department of Biochemistry and Biophysics, University of Pennsylvania, Philadelphia, PA 19104, USA
[2] Department of Chemistry, University of Manchester, Manchester M139PL, United Kingdom

41. Colloquium Mosbach 1990
The Molecular Basis of Bacterial Metabolism

determine how factors contributing to cofactor solvation at these sites control electron transfer performance.

Here, we present results of these investigations at the $Q_A$ and $Q_B$ sites, with emphasis on comparisons of the properties of the Q and $Q^-$ interactions at the two sites, from which an understanding of the general features promoting quinone-mediated redox catalyses is emerging.

## 2 Results and Discussion

### *2.1 Contributions to Binding Specificity at the $Q_A$ and $Q_B$ Sites*

Comparison of the $Q_A$ site $\Delta G_B^0$ values of an extensive number of systematically substituted quinone and analog compounds has led to a free energetic view of the head group binding region as 11–14 Å long in the direction perpendicular to the carbonyl axis, with low tolerance for extension normal to the ring plane and a set of steric features which block cofactor extension in the ring plane (Gunner et al. 1985, 1990). As shown in Fig. 1, a single quinone carbonyl oxygen atom dominates hydrogen bond contact with protein ($\Delta H < -3.5$ kcal/mol), acting to fix the orientation of the ring system in the site. In general, a variety of heteroatom

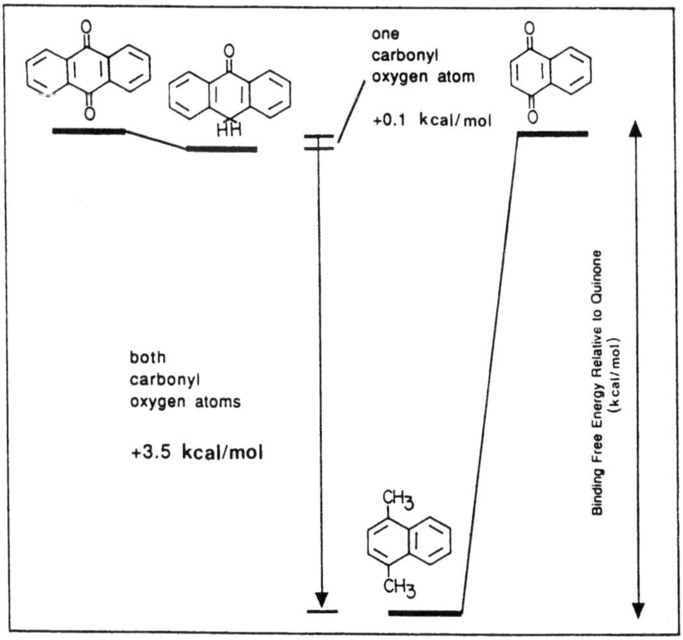


**Fig. 1.** Effect of quinone carbonyl oxygen atom removal on binding free energy at the $Q_A$ site in hexane solution. Values refer to the intrinsic interaction enthalpy loss for each pair of comparison compounds (AQ/anthrone and NQ/1,4-dimethylnaphthalene) estimated based on the treatment of Jencks (1975). Since hexane solvates ligands only through weak dispersion forces, comparison of the $\Delta G_B^0$ values determined in this solvent can directly reveal the presence of hydrogen bond interactions

substituents (e.g., -OH, $\Delta H = -2.3$ kcal/mol; -CHO, -2.0; -NH2, -1.4) are capable of specific, hydrogen bonded interaction with the site. Figure 2 shows the effects of isoprene tail length on UQ-$Q_A$ site $\Delta G^0_B$. The loss of tail hydrophobic contributions to $\Delta G^0_B$ in water after addition of the third isoprene unit suggests that the first three units are harbored within the protein. Energetically unfavorable intrachain torsion potentials in the bound tail are suggested to attenuate a net favorable tail-site interaction in the binding region (Warncke et al. 1990). Nevertheless, Fig. 3 shows that the $Q_A$ site interacts far more weakly with saturated tail structures compared with the isoprene tail, and thus that a specificity is exhibited for the native structure.

An identical approach has been pursued at the $Q_B$ site. By using low potential $Q_A$ site replacements to increase the equilibrium constant for $Q_A$- to $Q_B$ electron transfer, the capability of the $Q_B$ site to interact (and function) with a variety of benzo- and naphtho-quinones (BQ and NQ, respectively), including the 1,2-BQ and NQ isomers, has been revealed (Giangiacomo and Dutton 1989). The aqueous phase affinities of parent structures, as well as the magnitude of methyl and fused aromatic ring substitution-induced changes in BQ affinity, are attenuated relative to the $Q_A$ site (e.g., $\Delta\Delta G^0_B$ for trimethyl-BQ minus BQ is -3.4 and -0.6 kcal/mol at the $Q_A$ and $Q_B$ sites in water, respectively). The difference in head group substituent effects at the $Q_B$ versus the $Q_A$ site may be in part due to the increased polarity of the amino acid side chain environment in the $Q_B$ site (Michel et al. 1986; Allen et al. 1987), or compensating interactions from the substantial quantity of in situ water observed in the $Q_B$ site of *Rp. viridis* (Deisenhofer and Michel 1989).

In contrast, the patterns of tail-induced changes in quinone-$Q_B$ site affinity shown in Fig. 2, and dominance of a single carbonyl group in the interaction, suggested by the minor (0.3 kcal/mol) change in the $\Delta G^0_B$ for NQ accompanying rearrangement from the 1,4- to the 1,2-configuration, are comparable with features of the $Q_A$ site interactions.

The free energetic aspects of the sites revealed by the binding studies are well correlated with protein structural features observed in the X-ray crystallographic structures of the RC from *Rhodopseudomonas viridis* (Michel et al. 1986) and *Rb. sphaeroides* (Chang et al. 1986; Allen et al. 1987). There is also some resemblance between binding determinants at the bacterial sites and that of the PS II RC derived from quantitative structure-activity analyses (QSAR) (Trebst et al. 1984).

### 2.2 Protein Control of Cofactor Electrochemistry

Protein control of quinone electrochemistry at the catalytic sites is exercised through the selective solvation of the redox species. For $Q^-$ and Q bound at the $Q_A$ and $Q_B$ sites, the difference in the protein solvation free energies contributing to $E_m$ (in situ) ($\Delta\Delta G^0_{sol}$) can be resolved from the contribution of the intrinsic cofactor electrochemical potential [as given by the gas phase free energy of electron attachment ($\Delta G^0_a$), which is approximately equal to the electron affinity, EA] through application of the following equation commonly used to estimate $\Delta\Delta G^0_{sol}$ for organic redox couples in solution (Heinis et al. 1988; Shalev and Evans 1989):

$$\Delta\Delta G^0_{sol}(Q^- - Q) = EA + 23.06[E_m(\text{in situ})] - 4.79, \quad (1)$$

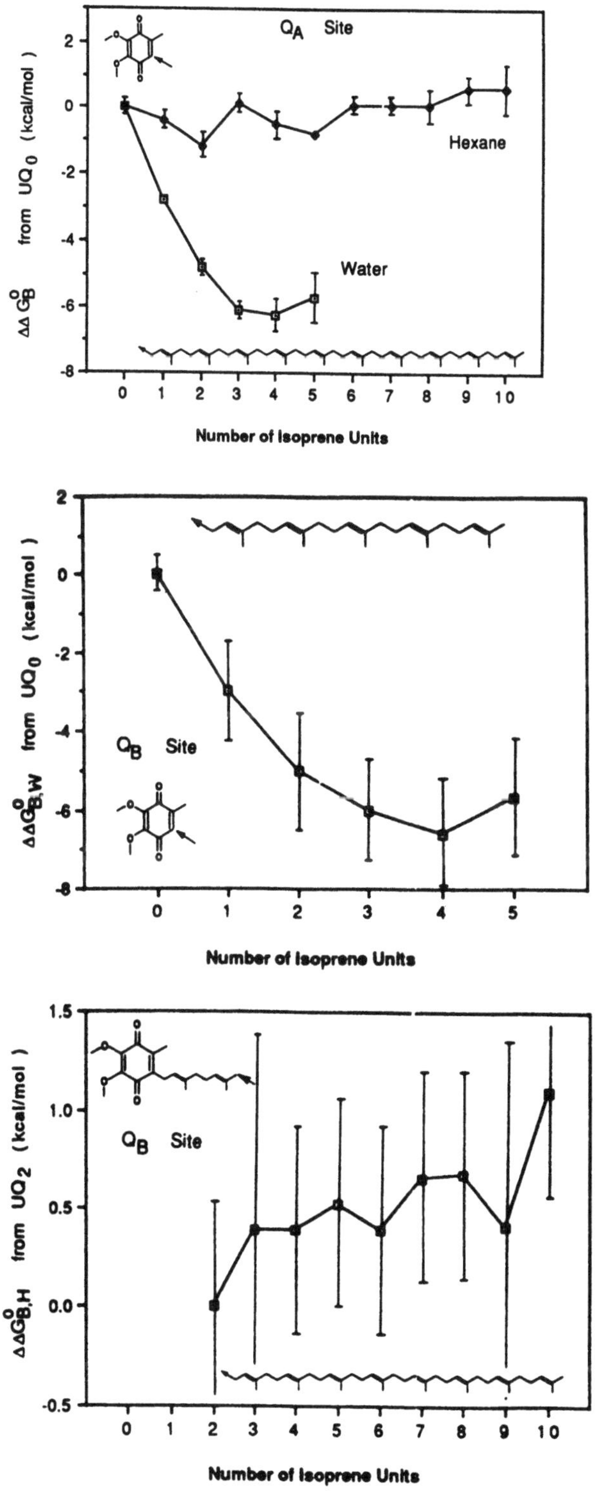


**Fig. 2.** Changes in binding free energy of $UQ_o$ at the $Q_A$ and $Q_B$ sites in aqueous and hexane solution as a function of isoprene tail length. Due to technical limitations, the contributions of the first two isoprene units to $Q_B$ site $\Delta G^0_B$ could not be obtained. A weak dependence of quinone affinity on isoprene tail length has also been observed in micellar systems (McComb et al. 1989)

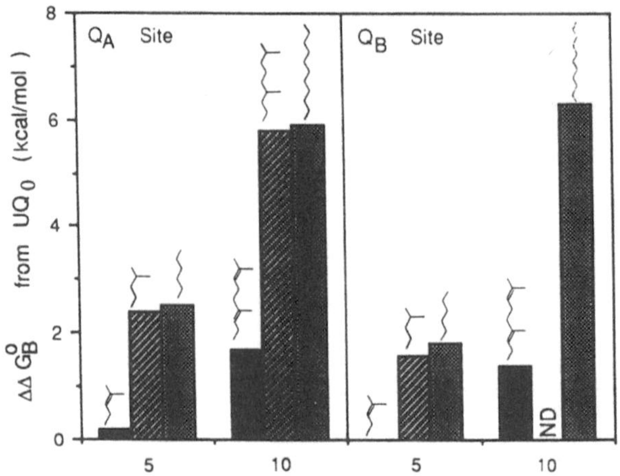


**Fig. 3.** Changes in binding free energy of $UQ_0$ at the $Q_A$ and $Q_B$ sites in hexane solution induced by substitution with various five- and ten-carbon atom tail structures. The site displays a strong specificity for interaction with the isoprene relative to saturated alkyl structures

where the units of $\Delta\Delta G^0_{sol}$ are given in kcal/mol and EA and $E_m$ (in situ) in V. Values of $E_m$ (in situ) for the $Q_A$ and $Q_B$ sites are from Woodbury et al. (1985) and Giangiacomo et al. (1990), respectively.

Figure 4 shows estimated values of $\Delta\Delta G^0_{sol}$ at the $Q_A$ and $Q_B$ sites, as well as in dimethylformamide (DMF) solution, as a function of the cofactor EA value. The magnitudes of the selective solvation of the semiquinone species at each site are comparable and of large magnitude, even in relation to the high polarity organic solvent DMF. The overlap of the $Q_A$ and $Q_B$ site data suggests that similar factors are involved in the determination of $\Delta\Delta G^0_{sol}$ at each site, which include: (1) redox state-dependent changes in quinonoid oxygen atom lone-pair basicity, which contribute to the localized, hydrogen bonding interactions with the sites (hydrogen bond contacts between the semiquinone and the sites are implicated by ENDOR studies; Feher et al. 1985). Methods recently devised to determine microscopic contributions to quinone solvation (Keske et al. 1990) suggest that up to −10 kcal/mol of the differential solvation free energy can be supplied by these interactions. The increase in $\Delta\Delta G^0_{sol}$ at the $Q_B$ site for EA $>$ 1.9 eV suggests that protein solvation of the reduced species maybe sensitive to the charge density localized on the quinonoid oxygen atoms, which has been shown to diminish with increasing EA value (Heinis et al. 1988). (2) Delocalized electrostatic interactions with the protein environment surrounding the sites (e.g., see Warshel 1981) apparently make a major contribution to the large magnitude of $\Delta\Delta G^0_{sol}$.

Included in Fig. 4 are two exotic $Q_A$ site cofactors: (1) tetrafluorofluorenone (TFF), used to assess the effect of removal of a single quinonoid oxygen atom on the functional interaction with the site, and (2) *meta*-dinitrobenzene (DNB), which represents a complete departure from the quinonoid structure. The observation

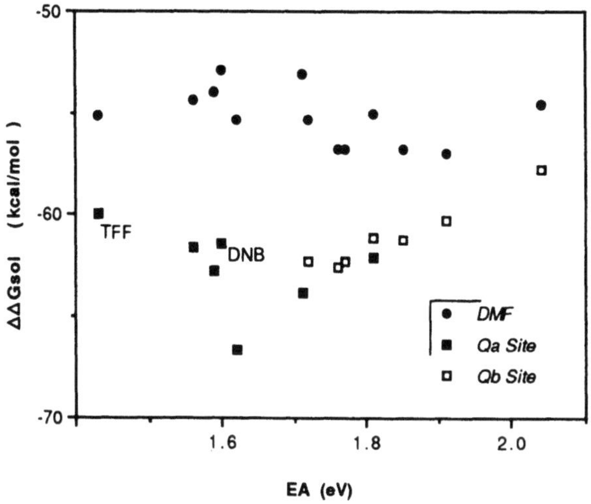


**Fig. 4.** Estimated values of $\Delta\Delta G^0_{sol}$ at the $Q_A$ and $Q_B$ sites and in DMF solution for quinone and exotic cofactors of varying EA value. Values of $\Delta\Delta G^0_{sol}$ are estimated as described in the text. EA values are from Heinis et al. (1988) and Shalev and Evans (1989). The EA value for TFF is estimated from the $E_m$ (DMF)-EA dependence exhibited by quinones (Heinis et al. 1988)

that the exotic cofactors, which lack key elements of the quinonoid structure, display $\Delta\Delta G^0_{sol}$ values at the $Q_A$ site comparable with quinones emphasizes the importance of delocalized dipolar interactions in protein determination of $E_m$ (in situ).

### 2.3 *Effects of Cofactor and Protein Structure on Electron Transfer Performance at the $Q_A$ Site*

Figures 5 and 6 present results from two strategies recently developed to determine the role of system structure in the performance of the charge-separating electron transfer from reduced bacteriopheophytin ($BPh^-$) to $Q_A$. Figure 5 shows a comparison of the electron transfer rate constants ($k_1$) for three exotic $Q_A$ site cofactors (K. Warncke, unpublished) at the representative temperature of 35 K with that of various quinone $Q_A$ site replacements (Gunner and Dutton 1989). The $k_1$ values of the exotic and quinone cofactors display a comparable dependence on $\Delta G^0_{et}$, which is also observed at temperatures from 14 to 298K. Figure 6 shows that values of the quantum yield for charge separation supported by various quinone $Q_A$ site replacements in the RC proteins from *Rp. viridis* (J. Keske and P.L. Dutton, unpublished) and *Rb. sphaeroides* (Gunner and Dutton 1989) also exhibit a smooth dependence on $\Delta G^0_{et}$, despite significant alterations in the RC protein primary structure between the two species (Komiya et al. 1988).

These results show that, whether from the perspective of the $Q_A$ site cofactor or the protein, dramatic alterations of the system structure do not lead to significant deviations from the dependence of kinetic performance on reaction free energy demonstrated or expected from previous examination of these reactions in the

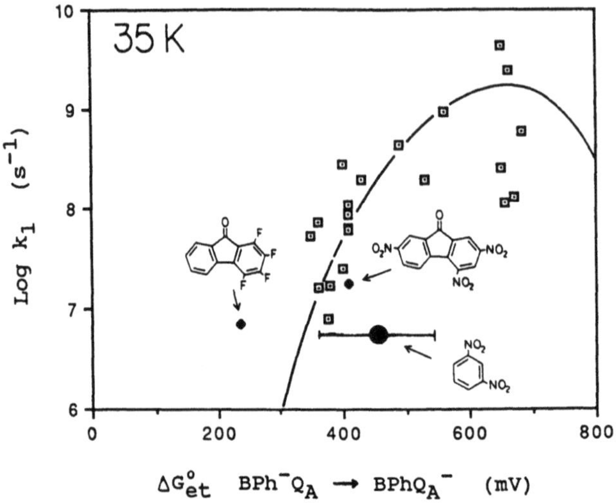


**Fig. 5.** Dependence of log $k_1$ on $\Delta G^0_{et}$ for quinone and exotic cofactors at 35 K. Values of $k_1$ were determined from the yield of the $[BChl]_2^+$ $Q_A^-$ state produced by xenon flash illumination of Q-bound RC samples in quartz EPR tubes, monitored at the peak of the g = 2.0026 EPR signal, as described in Gunner and Dutton (1989). The *solid line* represents the fit of the quinone data (including the temperature dependence of $k_1$ for T = 14,35,113, and 295 K) to a theoretical rate expression (Gunner and Dutton 1989). The current standard error in the determination of $E_m$ (in situ) for *m*-DNB is shown

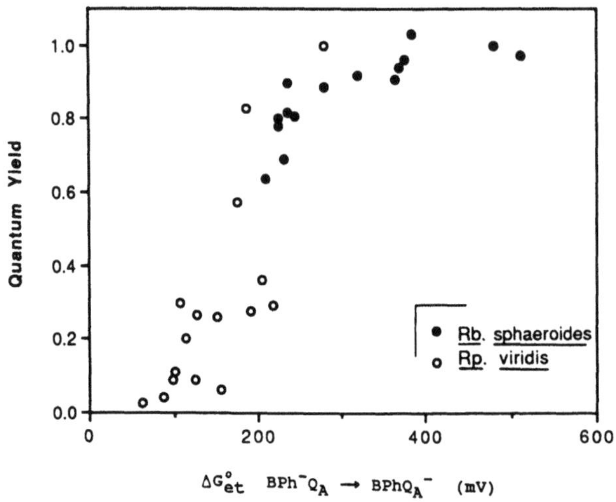


**Fig. 6.** Dependence of the quantum yield for charge separation on $\Delta G^0_{et}$ for various quinone $Q_A$ site replacements in the reaction centers from *Rb. sphaeroides* and *Rp. viridis*. The data for *Rb. sphaeroides* are from Gunner and Dutton (1989). The data for menaquinone-9 depleted RC protein from *Rp. viridis* (Keske and Dutton, unpublished) were obtained using methods described in Gao et al. (1990)

native *Rb. sphaeroides* system (Gunner and Dutton 1989). The continuity of the dependence of rate on reaction free energy shown in Figs. 5 and 6 suggests that fundamental mechanistic features involved in the $BPh^- \rightarrow Q_A$ electron transfer are not significantly altered.

## 3 General Conclusions Regarding Quinone-Mediated Redox Catalysis

Insights regarding the following points, gained from comparisons of the $Q_A$ and $Q_B$ site results, illuminate what appear to be general features of quinone-mediated redox catalyses:

1. *Functional specificity*: In view of the demonstrated ability of each site to mediate electron transfers using a variety of molecular structures, the native polyisoprene tail substituent is seen to play a critical role in the in vivo specificity of the quinone-site interaction by: (a) increasing the effective concentration of the quinone in the membrane interior relative to the aqueous medium (an "entropic enhancement" to the Q-mediated redox catalyses driven by the hydrophobic effect; Jencks 1975), and (b) utilization of a further "specificity filter" mechanism, involving interaction with the protein, which discriminates against competitive interference by membrane-soluble amphiphiles.

2. *Selective solvation of the Q- and Q species*: The estimated $\Delta\Delta G^0_{sol}$ values presented in Fig. 4 clearly demonstrate, as inferred from calculated $K_D$ values for Q- (Warncke and Dutton 1989; Wraight and Shopes 1989; Giangiacomo et al. 1990) that the environment surrounding the quinone catalytic sites is engineered for extremely strong interaction with the reduced cofactor species.

3. *Kinetic performance*: The results presented in Figs. 5 and 6 indicate that the value of $\Delta G^0_{et}$ dominates the determination of the rates of the $BPh^- \rightarrow Q_A$ electron transfer. Thus, a vital component of the control of electron transfer performance by the sites is simply the establishment of cofactor electrochemistry.

These results indicate that the definition of the functional properties of a particular quinone redox catalytic site can be determined in large part by the distance from, as well as the electrochemical nature of, redox centers in its vicinity.

## References

Allen, J.P., Feher, G., Yeates, T.O., Komiya, H. & Rees, D.C. (1987) Structure of the reaction center from *Rhodobacter sphaeroides*: the cofactors. Proc. Natl. Acad. Sci. USA 84:5730–5734

Braun, B.S., Benbow, U., Lloyd-Williams, P., Bruce, J.M. & Dutton, P.L. (1986) Determination of partition coefficients of quinones by high-performance liquid chromatography. Meth. Enzymol. 125:119–129

Chang, C.H., Tiede, D.M., Tang, J., Smith, U., Norris, J. & Schiffer, M. (1986) Structure of *Rhodopseudomonas sphaeroides* R-26 reaction center. FEBS Lett. 205:82–86

Deisenhofer, J. & Michel, H. (1989) The photosynthetic reaction center from *Rhodopseudomonas viridis*. Biophys. J. 55:1a

Dutton, P.L. (1986) Energy transduction in anoxygenic photosynthesis. In: Staehelin, A. & Arntzen, C.J. (eds) Encyclopedia of plant physiology: photosynthetic membranes. Springer, Berlin Heidelberg New York, pp. 197–237

Feher, G., Isaacson, R.A., Okamura, M.Y. & Lubitz, W. (1985) ENDOR of semiquinones in RCs from *Rhodopseudomonas sphaeroides*. In: Michel-Beyerle, M.E. (ed) Antennas and reaction centers of photosynthetic bacteria. Springer, Berlin Heidelberg New York, pp.174–189

Gao, J.L., Shopes, R.J. & Wraight, C.A. (1990) Charge recombination between oxidized high-potential c-type cytochromes and $Q_A^-$ in reaction centers from *Rhodopseudomonas viridis*. Biochim. Biophys. Acta (in press)

Giangiacomo, K.M. & Dutton, P.L. (1989) In photosynthetic reaction centers, the free energy difference for electron transfer between quinones bound at the primary and secondary quinone-binding sites governs the observed secondary site specificity. Proc. Natl. Acad. Sci. USA 86:2658–2662

Giangiacomo, K.M., Gunner, M.R., Bruce, J.M. & Dutton, P.L. (1990) Quinone and semiquinone binding free energies and electrochemistry in the $Q_B$ site of the reaction center protein from *Rhodobacter sphaeroides*. (in preparation)

Gunner, M.R. & Dutton, P.L. (1989) Temperature and free energy dependence of the electron transfer from $BPh^-$ to $Q_A$ in reaction center protein from *Rhodobacter sphaeroides* with different quinones as $Q_A$. J. Am. Chem. Soc. 111:3400–3412

Gunner, M.R., Braun, B.S., Bruce, J.M. & Dutton, P.L. (1985) The characterization of the $Q_A$ binding site of the reaction center of *Rhodopseudomonas sphaeroides*. In: Michel-Beyerle, M.E. (ed) Antennas and reaction centers of photosynthetic bacteria. Springer, Berlin Heidelberg New York, pp. 298–304

Gunner, M.R., Warncke, K., Braun, B.S., Bruce, J.M. & Dutton, P.L. (1990) The affinity of quinones for the $Q_A$ site of the reaction center protein from *Rhodobacter sphaeroides*. (in preparation)

Hauska, G. & Hurt, E. (1982) Pool function behavior and mobility of isoprenoid quinones. In: Trumpower, B.L. (ed) Function of quinones in energy conserving systems. Academic Press, New York, pp. 87–110

Heinis, T., Chowdhury, S., Scott, S.L. & Kebarle, P. (1988) Electron affinities of benzo-, naphtho-, and anthraquinones determined from gas-phase equilibria measurements. J. Am. Chem. Soc. 110:400–407

Jencks, W.P. (1975) Binding energy, specificity and enzymic catalysis – the Circe Effect. Adv. Enzymol. 43:219–410

Keske, J.M., Bruce, J.M. & Dutton, P.L. (1990) Assessment of organic solute-water hydrogen bond strengths by analysis of multiple solvent partitioning data. Z. Naturforsch. (in press)

Komiya, H., Yeates, T.O., Rees, D.C., Allen, J.P. & Feher, G. (1988) Structure of the reaction center from *Rhodobacter sphaeroides* R-26 and 2.4.1: symmetry relations and sequence comparisons between different species. Proc. Natl. Acad. Sci. USA 85:9012–9016

McComb, J.C., Stein, R.R. & Wraight, C.A. (1989) Investigations on the influence of headgroup substitution and isoprene side-chain length in the function of primary and secondary quinones of bacterial reaction centers. Biochim. Biophys. Acta 1015:156–171

Michel, H. Epp, O. & Deisenhofer, J. (1986) Pigment-protein interactions in the photosynthetic reaction centre from *Rhodopseudomonas viridis*. EMBO J. 5:2446–2451

Okamura, M.Y., Isaacson, R.A. & Feher, G. (1975) Primary acceptor in bacterial photosynthesis: obligatory role of ubiquinone in photoactive reaction centers of Rhodopseudomonas sphaeroides. Proc. Natl. Acad. Sci. USA 72:3491–3495

Prince, R.C., Gunner, M.R. & Dutton, P.L. (1982) Quinones of of value to electron transfer studies: oxidation-reduction potentials of the first reduction step in an aprotic solvent. In: Trumpower, B.L. (ed) Function of quinones in energy conserving systems. Academic Press, New York, pp. 29–34

Shalev, H. & Evans, D.H. (1989) Solvation of anion radicals: gas phase vs. solution. J. Am. Chem. Soc. 111:2667–2674

Trebst, A., Donner, W. & Draber, W. (1984) Structure activity correlation of herbicides affecting plastoquinone reduction by photosystem II: electron density distribution in inhibitors and plastoquinone species. Z. Naturforsch. 39c:405–411

Trumpower, B.L. (1982) Function of quinones in energy conserving systems. Academic Press, New York

Warncke, K. & Dutton, P.L. (1989) Quinone electrochemistry and reaction center $Q_A$ site affinity addressed in a common aprotic solvent system. Biophys. J. 55:181a

Warncke, K., Gunner, M.R., Braun, B.S., Gu, L., Yu, C. A. & Dutton, P.L. (1990) Effect of hydrocarbon tail structure on the interaction of quinones with the $Q_A$ and $Q_B$ sites of the reaction center protein from *Rhodobacter sphaeroides*. (in preparation)

Warshel, A. (1981) Calculations of enzyme reactions: calculations of pKa, proton transfer reactions, and general acid catalysis reactions in enzymes. Biochemistry 20:3167–3177

Woodbury, N.W., Parson, W.W., Gunner, M.R., Prince, R.C. & Dutton, P.L. (1985) Radical pair energetics and decay mechanisms in reaction centers containing anthraquinones, naphthoquinones, or benzoquinones in place of ubiquinone. Biochim. Biophys. Acta 851:6–22

Wraight, C.A. (1982) The involvement of stable semiquinones in the two-electron gates of plant and bacterial systems. In: Trumpower, B.L. (ed) Function of quinones in energy conserving systems. Academic Press, New York, pp. 181–198

Wraight, C.A. & Shopes, R.J. (1989) Quinone binding and herbicide activity in the acceptor quinone complex of bacterial reaction centers. In: Barber, J. & Malkin, R. (eds) Techniques and new developments in photosynthesis research. Plenum, New York, pp. 183–190

# Relationship of Cytochrome caa$_3$ from *Thermus thermophilus* to Other Heme- and Copper-Containing Terminal Oxidases

M.W. Mather, P. Springer, and J.A. Fee[1]

## 1 Introduction

Cytochrome oxidases are a key component of the energy metabolism of most aerobic organisms from mammals to bacteria. They are the final enzyme of the membrane-associated respiratory chain responsible for converting the chemical energy of reduced substrates to a transmembrane electrochemical potential, which is used by the cell for a wide variety of energy-requiring processes. The most widely studied oxidase is the cytochrome c oxidase (cytochrome aa$_3$ oxidase) of the mammalian mitochondrion. This complex, integral membrane protein contains 13 subunits and 4 canonical metal centers: heme centers, a and a$_3$; copper centers, Cu$_A$ and Cu$_B$. It is responsible for electron transfer from reduced cytochrome c to dioxygen with the concomitant reduction of dioxygen to water and the coupled vectorial transfer of protons across the mitochondrial membrane (see Chan and Li 1990; Palmer 1987 for recent reviews).

Bacterial membranes have been shown in many cases to contain respiratory components very similar to those of the mitochondrion. In particular, the bacterial cytochrome aa$_3$ oxidases were found to possess similar spectroscopic and functional properties to the mitochondrial enzyme, but a much simpler subunit structure. Purified, active bacterial oxidases contain the four canonical metal centers, but consist of only one to three subunits (reviewed in Anraku 1988). In the case of *Paracoccus denitrificans* (Raitio et al. 1987) these three subunits were shown to be homologous to the three largest subunits of the eukaryotic oxidase, which are synthesized in the mitochondrion (the remaining subunits being imported from the cytoplasm). From recent results, it is now apparent that there are at least three types of heme- and copper-containing terminal oxidases: the classical aa$_3$ type discussed above; the recently discovered ba$_3$ type, in which the heme A of cytochrome a is replaced by protoporphyrin IX (heme B) (Zimmermann et al. 1988); and the bacterial cytochrome bo oxidase (a quinol oxidase), containing two B hemes in place of the two A hemes of the aa$_3$ type and at least one copper with properties resembling Cu$_B$ (Kita et al. 1984; Hata et al. 1985; Salerno et al. 1989). The cytochrome bo oxidase from *E. coli*, like the aa$_3$ oxidase from *P. denitrificans*, has now been shown to contain subunits homologous to the three largest subunits of the mitochondrial oxidase (Chepuri et al. 1989). The sequence of the single subunit cytochrome ba$_3$ is not yet available, but, because the metal centers reside in similar environments (Zimmermann et al. 1988; Einarsdóttir et al. 1989), it seems

[1] Biochemistry Section and Stable Isotope Resource, Isotope and Structural Chemistry Group, Los Alamos, National Laboratory, Los Alamos, NM 87545, USA

41. Colloquium Mosbach 1990
The Molecular Basis of Bacterial Metabolism

likely that it too will contain homologies to one or more of the mitochondrial subunits. The recently isolated heme A- and copper-containing oxidase from thermophilic archaebacteria (Annemüller and Schäfer 1989), which appears to have only one subunit, may prove to be yet another subtype of this superfamily of oxidases.

The thermophilic eubacterium *Thermus thermophilus* has proven to be a good source of stable, comparatively simple respiratory enzymes (reviewed in Fee et al. 1986). Under appropriate growth conditions, the membrane fraction contains significant amounts of two terminal oxidases, cytochrome $caa_3$ and cytochrome $ba_3$. The latter oxidase is the subject of ongoing and future studies in our laboratory. This communication will principally be concerned with recent results from our studies of *Thermus*' cytochrome $caa_3$ oxidase.

Cytochrome $caa_3$ has been isolated as a two subunit (55 and 33 kDa) enzyme containing the four canonical metal centers and an additional C heme (Hon-nami and Oshima 1984; Yoshida et al. 1984). The enzyme has been extensively characterized and shown to be essentially identical spectroscopically to a combination of cytochrome $aa_3$ plus cytochrome c (see Fee et al. 1986 and references cited therein). The $caa_3$ oxidase has been shown to function as a redox-coupled proton pump when reconstituted in proteoliposomes (Sone et al. 1983; Yoshida and Fee 1984). The finding that the 33 kDa subunit, when resolved from the $caa_3$ complex and separately purified, bound the C heme but apparently contained no A heme or copper was a principal factor leading to the hypothesis that the 55 kDa subunit (presumably homologous to subunit I of the mitochondrial oxidases) contains the four canonical metal centers and constitutes the catalytic core of the oxidase (Yoshida et al. 1984; see also Fee et al. 1986). This hypothesis conflicts with the more generally accepted view that subunit I contains the two A hemes and $Cu_B$, but that $Cu_A$ is bound to subunit II. The latter hypothesis is consistent with a large number of spectroscopic, copper depletion, and other experiments (see Holm et al. 1987; Palmer 1987; Chan and Li 1990), but may be inconsistent with the isolation of the single subunit cytochrome $ba_3$ oxidase (Zimmermann et al. 1988) and the preparation of an enzymatically digested cytochrome $aa_3$ oxidase from *P. denitrificans* which consists of the major portion of subunit I only, yet retains activity and the optical spectrum of the native enzyme (Müller et al. 1988a,b). In this communication we will describe preliminary results of DNA sequencing experiments with the cytochrome $caa_3$ oxidase, initially undertaken to determine the nature of the subunits of this oxidase and shed light on the distribution of the metal centers. We will speculate on oxidase gene and protein structures and evolutionary relationships in the light of these results and recent sequencing results from other groups.

## 2 Experimental Procedures

The procedures used to clone the oxidase genes from *T. thermophilus* have been described (Fee et al. 1988; Mather 1988). DNA sequencing was performed using the dideoxy chain termination method (Sanger et al. 1977). Several methods were used to overcome polymerase terminations, gel compressions, and other problems caused by the high guanine/cytosine content of *Thermus* DNA, including the use of 7-deaza-deoxyguanosine and deoxyinosine 5′-triphosphates in the sequencing reactions; details will be published elsewhere (Mather et al. 1990). Computer sequence analyses were performed using the University of Wisconsin GCG programs (Devereux et al. 1984) and additional software written by one of the authors (MWM).

## 3 Results and Discussion

### *3.1 Summary of Results*

We have cloned and sequenced a ~4 kb region of the *Thermus thermophilus* chromosome containing genes encoding subunits of the cytochrome $caa_3$ oxidase. Figure 1 is a schematic representation of this portion of the chromosome. A few small regions of uncertainty in the DNA sequence remain to be clarified by additional sequencing experiments, most notably, the region following the second *Hin*dIII site and presumably containing the end of subunit I and the start of subunit III. Comparative analyses of the deduced amino acid sequences of these subunits together with those of other available oxidase sequences lead to a number of conclusions:

1. *T. thermophilus* contains genes for three oxidase subunits: I, IIc, and III. These genes are contiguous on the *T. thermophilus* chromosome.
2. The sequences of the *T. thermophilus* oxidase subunits are very similar to the corresponding sequences of all other oxidase subunits, I, II, and III.
3. Hydrophobicity analyses show that all oxidase subunits I, II, and III which have been sequenced have hydrophobic regions (putative transmembrane helices) at corresponding positions. This, together with the sequence and spectroscopic similarities, suggests a basic structural similarity for all the heme/copper oxidases from mitochondrial and eubacterial sources.[2]

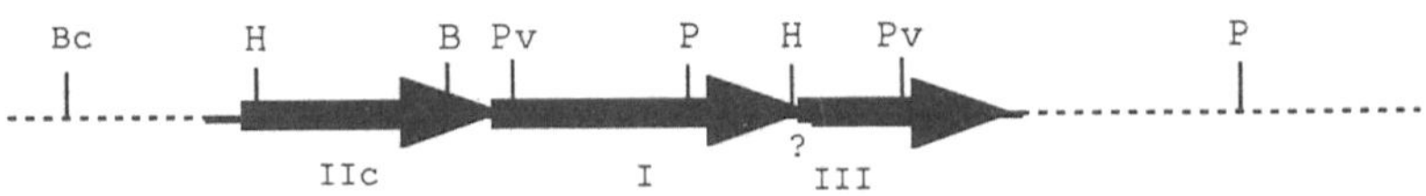


**Fig. 1.** Map of the region of the *T. thermophilus* chromosome containing the genes encoding the subunits of the cytochrome $caa_3$ oxidase. The *solid line* indicates the region for which the DNA sequence has been obtained. *Arrows*, labeled underneath as *IIc*, *I*, or *III*, indicate the positions of the genes for the corresponding subunits of the oxidase. *Vertical marks* indicate the positions of restriction endonuclease sites: *B Bam*HI; *Bc Bcl*I; *H Hind*III; *P Pst*I; and *Pv Pvu*II. The *question mark* indicates the region of current ambiguity in the DNA sequence near the end of the subunit I gene/beginning of the subunit III gene

[2] Sweet and Eisenberg (1983) have pointed out the basic correlation between structural elements and hydrophobicity patterns. Thus, the probability of structural similarity does not depend solely on any specific structural interpretation of the hydrophobicity plot, such as the existence of transmembrane helices formed from extended segments of hydrophobic amino acids.

4. Subunit III of the *T. thermophilus* oxidase is apparently lost during purification, as in the case of *P. denitrificans* (Saraste et al. 1986).
5. An acidic cytochrome c domain is fused to the carboxy-terminus of the *T. thermophilus* subunit II. This may represent a fusion of substrate (cytochrome c) to enzyme (cytochrome c oxidase).
6. The two cysteine and two histidine residues proposed to coordinate $Cu_A$ in subunit II (Holm et al. 1987) are retained in *T. thermophilus* subunit IIc, as in all other cytochrome $aa_3$ oxidases. The sequences near the putative ligands are highly conserved. *E. coli* bo oxidase subunit II, in contrast, lacks the proposed $Cu_A$ liganding residues, but otherwise retains a substantial degree of sequence similarity in this region of the polypeptide.
7. Subunit I is the most highly conserved subunit. Comparison of all presently known subunit I sequences indicates that only seven histidines are conserved. According to the results of spectroscopic experiments, these probably constitute a near minimal set of ligands for the metal cofactors found in subunit I: cytochrome a (2 His), cytochrome $a_3$ (1 His), $Cu_B$ (2–3 His).
8. Comparison of the order and extent of the bacterial oxidase genes among several species suggests that genomic rearrangements involving the subunit III gene have occurred.

In the remainder of this communication we discuss some specific results regarding three interesting points: (1) The overall similarity of subunits I, II, and III among all the oxidases; (2) some possible genomic variations which may result in variations in the position of the amino terminal portion of subunit III among the bacterial oxidases; and (3) the conservation of sequence near the proposed $Cu_A$ coordinating residues in subunit II.

### 3.2 Similarity of Oxidase Sequences and Structures

We have prepared alignments of all available sequences of oxidase subunits I, II, and III using the multiple alignment algorithm of Gribskov et al. (1987) with some minor adjustments made upon consideration of conserved amino acid residues and relative locations of hydrophobic regions (which tend to be less well conserved and thus more difficult to align). The alignments will be published elsewhere. Here, we present a compilation of the degree of identity of subunits I, II, and III from some selected species (Table 1). The general rule of thumb is that a percent identity of 25% or more is good evidence of a significant match between two sequences, while scores below 15% provide no evidence of relatedness (Doolittle 1981, 1986). The strong sequence conservation of the oxidase subunits, especially of subunit I, among these evolutionarily distant organisms is readily apparent. In the case of subunits II and III of the *E. coli* bo oxidase, comparison to eukaryotic subunits II and III leave some doubt with regard to the significance of the postulated relationship, e.g., note the relatively low percent identities in the bovine/*E. coli* comparison of Table 1. Comparison to the subunits from *T. thermophilus* (Table 1) or *Bacillus* PS3 (not shown), however, clearly establishes the relationship of the cytochrome bo oxidase subunits to the corresponding cytochrome $aa_3$ oxidase subunits. The *T. thermophilus* subunits have the greatest similarity to those from

**Table 1.** Percent of identical amino acid residues[a] in cytochrome or quinol oxidase subunits I, II, and III from six species compared to the corresponding subunits from the bovine mitochondrial oxidase or from the *T. thermophilus* oxidase

Comparison to the bovine oxidase

| Source[b] | I | Subunit II / IIs[c] | III |
|---|---|---|---|
| *D. yakuba* | 75.5 | 59.8 / 65.3 | 65.0 |
| *S. cerevisiae* | 59.6 | 44.7 / 50.0 | 44.2 |
| *P. denitrificans* | 54.3 | 36.1 / 43.1 | 51.0 |
| *Bacillus* PS3 | 44.0 | 24.9 / 29.4 | 24.9[d] |
| *T. thermophilus* | 39.8 | 26.8 / 35.6 | 29.5 |
| *E. coli* | 40.1 | 17.7 / 22.5 | 23.4[d] |

Comparison to the *T. thermophilus* oxidase

| Source[b] | I | Subunit II / IIs[c] | III |
|---|---|---|---|
| Bovine | 39.8 | 26.8 / 35.6 | 29.5 |
| *D. yakuba* | 40.8 | 24.6 / 33.7 | 28.2 |
| *S. cerevisiae* | 39.4 | 27.3 / 33.3 | 27.8 |
| *P. denitrificans* | 42.6 | 28.9 / 36.9 | 30.1 |
| *Bacillus* PS3 | 47.1 | 33.3 / 39.4 | 29.0[d] |
| *E. coli* | 38.2 | 31.0 / 33.7 | 29.2[d] |

[a] Percent identity was calculated from an alignment of all published sequences for each subunit (and prepublication sequences for *T. thermophilus*, *Bacillus* PS3, and *E. coli*) over the extent of the shortest sequence in the alignment for each subunit and not including the initial methionine (this corresponds to residues 3–512, 10–214, and 2–259 of bovine subunits I, II, and III, respectively). The alignments will be published elsewhere. Sequences from plant or protozoan mitochondrial oxidases which have not been corrected for RNA editing were not included in the alignments. The formula used was: %ID = 100 X (number of identical residues) / (length of sequence in common).

[b] References for the sequences used: Bovine (Anderson et al. 1982; Steffens and Buse 1979); *Drosophila yakuba* (Clary and Wolstenholme 1983a,b); *Saccharomyces cerevisiae* (Coruzzi and Tzagoloff 1979; Bonitz et al. 1980; Thalenfeld and Tzagoloff 1980); *Paracoccus denitrificans* (Raitio et al. 1987; Steinrücke et al. 1987); *Bacillus* PS3 (Sone et al. 1988; Ishizuka et al. 1989); *Thermus thermophilus* (Mather et al. 1990); and *Escherichia coli* (Chepuri et al. 1989).

[c] IIs refers to the hydrophilic portion of subunit II which constitutes approximately the carboxy-terminal three-fifths of the subunit (residues 91–214 in the bovine oxidase subunit II for purposes of calculating the percent identity).

[d] Subunits III from *Bacillus* PS3 and from *E. coli* are missing sequences corresponding to an amino terminal segment present in the other oxidase subunits III (however, see the text for possible alternate locations of this sequence). The percent identity for these two subunits III is calculated for a shorter sequence, corresponding to residues 66–259 of bovine subunit III.

the gram positive thermophile *Bacillus* PS3 (except in the case of subunit III); otherwise they are not appreciably more similar to the other bacterial oxidase subunits than to subunits from mitochondrial oxidases. This is consistent with the proposed evolutionary distance of *Thermus* from other bacteria (reviewed by Woese 1987).

A separate tabulation for the hydrophilic portion of subunit II is presented in Table 1 to illustrate the higher degree of conservation of this portion of the subunit. The hydrophobic segments in the amino terminal portion of subunit II may be less conserved because they function as a membrane anchor for the peptide (Bisson et al. 1982a), which apparently requires only that the overall hydrophobicity of the segments be maintained.

As noted above, hydrophobicity plots of aligned oxidase subunits show that these subunits, sequenced from a wide variety of species, contain hydrophobic regions in essentially identical positions in the sequence alignments. For example, compare the top two traces in Fig. 2 (see, for example, Wikström et al. 1985 for additional comparative hydrophobicity plots of oxidase subunits). It is likely that many of these hydrophobic segments are transmembrane helices, as was found for the reaction centers from photosynthetic bacteria (Michel et al. 1986). Further, this alignment of the hydrophobic regions, and alternating hydrophilic regions, predicts a basic structural similarity (see Footnote 1) among the corresponding individual subunits (I, II, and III) from all sources examined so far. It is very likely, then, that general conclusions drawn from studies of the bacterial oxidases are valid for all the heme/copper oxidases.

### 3.3 *Organization of Bacterial Genes Encoding Oxidase Subunits*

The oxidase genes of *T. thermophilus* are contiguous on the chromosome in the order IIc, I, and III[3] (Figs. 1 and 3). The same gene order is found in the case of *Bacillus* PS3 and *E. coli* (see Fig. 3; Chepuri et al. 1989; Ishizuka et al. 1989). However, the genes for subunits III from *Bacillus* PS3 and *E. coli* are shorter than those of the other species. The translated peptide sequences do not have a region which corresponds to the amino terminal portion of subunit III from other sources, including the first two (of seven) hydrophobic segments. As shown in Figs. 2 and 3, however, there is an extra sequence element at the end of the *E. coli* subunit I gene which translates to a peptide sequence containing two hydrophobic segments, and this peptide sequence can be aligned to the beginning of the subunit III sequence alignment. A similar sequence element occurs in the case of *Bacillus* PS3; it is found as a separate open reading frame (orf) between the genes encoding subunits I and III (Fig. 3; Ishizuka et al. 1989). This suggests that transfer of the initial portion of the gene for subunit III to a separate gene or to the end of the subunit I gene has occurred without disrupting the basic structure of the oxidase protein complex. Since the subunit III gene follows the subunit I gene on the chromosome of these

[3] The exact end of the subunit I gene and beginning of the subunit III gene in *T. thermophilus* is uncertain because we have not yet obtained high quality sequencing gels for a small region which apparently includes the end of subunit I and the start of subunit III (Fig. 1). So far we have not identified a stop codon following subunit I; this should be clarified shortly.

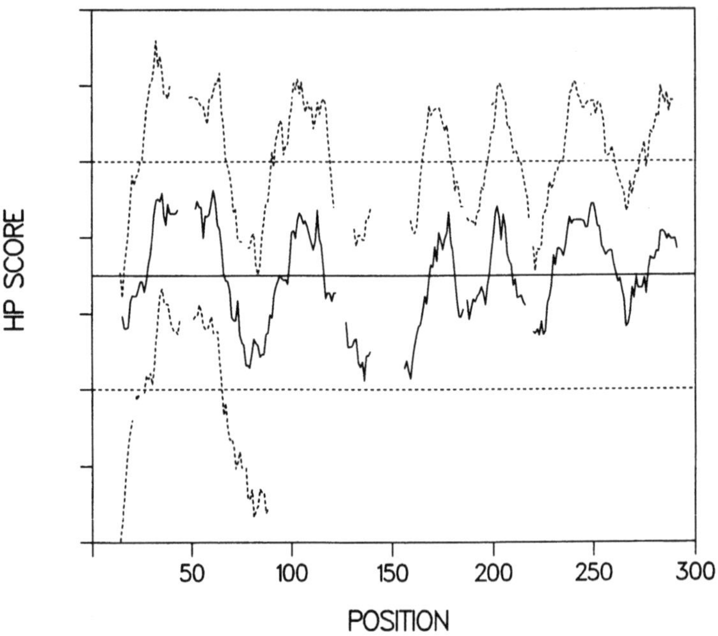


**Fig. 2.** Hydrophobicity plot of *T. thermophilus* (*upper dashed trace*) and bovine (*middle solid trace*) cytochrome oxidase subunits III together with the carboxy-terminal portion of *E. coli* cytochrome bo oxidase subunit I (*lower dashed trace*). The hydrophobicities were calculated using the MPH scale of Degli Esposti et al. (1989), and moving averages calculated with a window of 17 residues (Kyte and Doolittle 1982); the abscissa of the plot is taken from the position of the central residue of the window with respect to an alignment of subunit III sequences from 20 species (to be published elsewhere after finalization of the *T. thermophilus* sequence). The *marks* on the ordinate scale delineate 0.2 units on the hydrophobicity scale. The *horizontal lines* through the traces indicate the theoretical boundary between hydrophobic (*above* the line) and hydrophilic (*below* the line) averages. *Gaps* in the plot occur where one or more sequences have inserts relative to the sequence of the plotted species (e.g., the gap between the first two peaks of hydrophobicity occurring at about position 43 to 51 is due to additional residues present in the sequence of *P. denitrificans* subunit III which do not align with the other sequences). This type of hydrophobicity plot provides an assessment of whether the various hydrophobic and hydrophilic regions of each sequence occur in essentially the "same place". The bovine sequence was taken from Anderson et al. (1982), the *T. thermophilus* sequence represents data from our laboratory (Mather et al. 1990), and the *E. coli* sequence represents prepublication data graciously provided by Dr. R.B. Gennis (Chepuri et al. 1989). The similarity of the sequences in the region of the first two hydrophobic segments (positions 1 through 80) is relatively low; in particular, the alignment of the *E. coli* subunit I carboxy-terminal sequence is speculative (not significant by statistical tests)

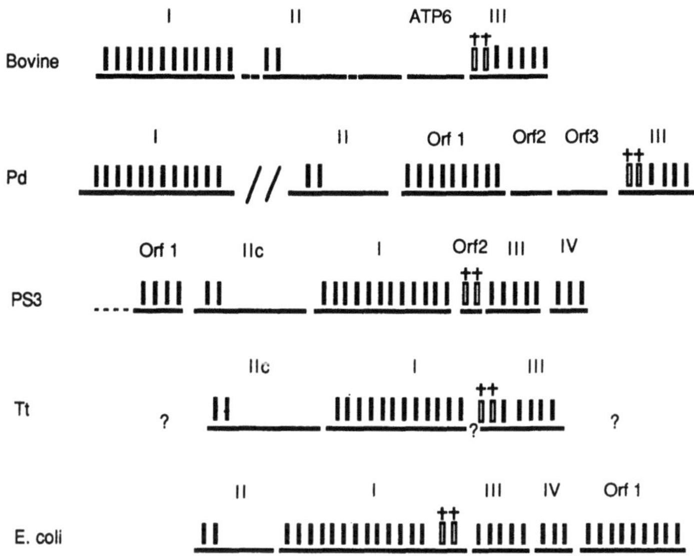


**Fig. 3.** Organization of the genes encoding oxidase subunits in several species. The relative positions of the oxidase genes are depicted for bovine mitochondrial (bovine), *Paracoccus denitrificans* (Pd), *Bacillus* PS3 (PS3), *Thermus thermophilus* (Tt), and *Escherichia coli* (*E. coli*) chromosomes. The gene encoding subunit I of the *P. denitrificans* oxidase is found in a separate location on the bacterial chromosome; this is indicated by *two slashes* in the figure. *Vertical bars* indicate the approximate positions of DNA coding sequence segments which translate to hydrophobic amino acid segments (putative membrane spanning helices); the *unfilled bars* (also marked above the bars by a †) denote the first two such hydrophobic segments present in all eukaryotic subunit III genes and peptides and their putative homologs in the bacterial oxidase genes. Genes labeled *I, II, III*, and *IV* in the various species appear to encode homologous protein subunits. *Orf1* (Orf = open reading frame) from *P. denitrificans* and from *E. coli*, and possibly from PS3, also appear to encode homologous peptides. "?" indicates regions where the DNA sequence was unknown or as yet uncertain. The PS3 and *E. coli* organization is based on DNA sequence data graciously provided prior to publication by Dr. N. Sone (Ishizuka et al. 1989) and Dr. R.B. Gennis (Chepuri et al. 1989). Data for bovine mitochondria were taken from Anderson et al. (1982) and for *P. denitrificans* from Raitio et al. (1987)

bacteria, it is easy to envision how this could be accomplished by mutations producing a stop codon at the appropriate place in the gene encoding subunit III, and, in the case of *E. coli*, eliminating a stop codon at the end of an originally shorter subunit I gene. We note that even more drastic genomic rearrangements were found in the case of bacterial operons encoding sugar phosphotransferase systems, which are also membrane-associated (reviewed in Robillard and Lolkema 1988). The nature of the junction region for the subunit I-subunit III genes in *T. thermophilus* is still under investigation (see Footnote 2).

### *3.4 Sequence Conservation in the Proposed Cu$_A$ Binding Region of Subunit II*

The most highly conserved region among subunits II of the cytochrome aa$_3$ oxidases is the region near the carboxy terminus which contains the two conserved histidines (161 and 204 of the bovine sequence) and two conserved cysteines (positions 196 and 200 of the bovine sequence) proposed to coordinate Cu$_A$ (see

Holm et al. 1987; Martin et al. 1988). It also contains a conserved methionine (bovine position 207) which was initially proposed to be a ligand to $Cu_A$ (Steffens and Buse 1979) and conserved aspartate and glutamate residues (bovine positions 158 and 198) thought to be involved in cytochrome c binding (Bisson et al. 1982b; Millet et al. 1983). The positions of these residues can be seen in Fig. 4. Since the *E. coli* cytochrome bo oxidase has been found to lack a $Cu_A$ center, while retaining two hemes and a $Cu_B$, it is interesting to note that the quinol oxidase subunit II does not retain any of the conserved residues currently proposed to coordinate $Cu_A$ or to participate in cytochrome c binding, yet has a high degree of overall similarity in this region to subunit IIc from the *T. thermophilus* $caa_3$ oxidase (37% identity; Fig. 4)[4]. This suggests to us that this region of amino acid sequence may form a unique three-dimensional structure which is conserved in all the heme/copper oxidases, in spite of the apparent loss of $Cu_A$ binding capacity in the bo oxidase. The *E. coli* cytochrome bo oxidase, then, has a homologous subunit II, but the proposed $Cu_A$ ligands are absent from this subunit; this result is consistent with, but does not substantiate, the proposal that the conserved cysteines and histidines in subunit II of the cytochrome $aa_3$ oxidases are ligands to $Cu_A$. In light of the hypothesis that $Cu_A$ is a site of proton pumping in cytochrome $aa_3$ oxidase (Gelles et al. 1986), it is also interesting to note that cytochrome bo oxidase has recently been found to translocate protons (Puustinen et al. 1989).

```
                   ** *          *    *
Tt   PAGVPVELEITSKDVIHSFWVPGLAGKRDAIPG
Ec   PANTPVYFKVTSNSVMNSFFIPRLGSQIYAMAG
                  x  x

                     * * * **  *  *
Tt   QTTRISFEPKEPGLYYGFCAELCGASHARMLFR
Ec   MQTRLHLIANEPGTYDGISASYSGPGFSGMKFK
                       x x x   x
```

**Fig. 4.** Comparison of the amino acid sequences of subunits II from *Thermus thermophilus* and *Escherichia coli* in the region containing the proposed ligands of $Cu_A$ (residues 130 through 195 in the *T. thermophilus* subunit II sequence, corresponding to residues 145 through 209 in the bovine subunit II sequence). Residues identical in the two sequences are shown as *outlined letters*. Residues totally conserved among all cytochrome $aa_3$ oxidases are indicated by *; conserved residues proposed to function as ligands to $Cu_A$ or to be involved in cytochrome c docking, and which are absent from the *E. coli* bo oxidase, are indicated by x (Taken from an alignment of bacterial oxidases to be published elsewhere)

*Acknowledgments.* We thank Professor G. Buse and his coworkers who originally supplied us with necessary amino acid sequences; Dr. N. Sone for providing us with the DNA sequence of the PS3 cytochrome $caa_3$ genes; Dr. R.B. Gennis who provided us with the DNA sequence of the *E. coli* bo oxidase genes in advance of publication; and the Life Sciences Division of LANL, especially Dr. M. Wilder, for making their VAX computers available and assisting us with their use. This work was supported by U.S.P.H.S. Grant GM35342 and performed under the auspices of the U.S. Department of Energy.

[4] The replacements of the conserved residues which are found in the *E. coli* subunit II are, however, in most cases evolutionarily conservative: Cys → Ser is the most frequent cysteine replacement; His → Asn, Asp → Ser and Glu → Ser are also common (Dayhoff et al. 1978).

## References

Anderson, S., de Bruijn, M.H.L., Coulson, A.R., Eperon, I.C., Sanger, F. & Young, I.G. (1982) Complete sequence of bovine mitochondrial DNA. Conserved features of the mammalian mitochondrial genome. J. Mol. Biol. 156:683–717

Annemüller, S. & Schäfer, G. (1989) Cytochrome aa$_3$ from the thermoacidophilic archaebacterium *Sulfolobus acidocaldarius*. FEBS Lett. 244:451–455

Anraku, Y. (1988) Bacterial electron transport chains. Annu. Rev. Biochem. 57:101–132

Bisson, R., Steffens, G.C.M. & Buse, G. (1982a) Localization of lipid binding domain(s) on subunit II of beef heart cytochrome c oxidase. J. Biol. Chem. 257:6716–6720

Bisson, R., Steffens, G.C.M., Capaldi, R.A. & Buse, G. (1982b) Mapping of the cytochrome c binding site on cytochrome c oxidase. FEBS Lett. 144:359–363

Bonitz, S.G., Coruzzi, G., Thalenfeld, B.E. & Tzagoloff, A. (1980) Assembly of the mitochondrial membrane system. Structure and nucleotide sequence of the gene coding for subunit I of yeast cytochrome oxidase. J. Biol. Chem. 24:11927–11941

Chan, S.I. & Li, P.M. (1990) Cytochrome c oxidase: understanding nature's design of a proton pump. Biochemistry 29:1–12

Chepuri, V., Lemieux, L., Au, D.C.T. & Gennis, R.B. (1990) The Sequence of the *cyo* Operon Indicates Substantial Structural Similarities between the Cytochrome *o* Ubiquinol Oxidase of *Escherichia coli* and the *aa*$_3$-type Family of Cytochrome *c* Oxidases. J. Biol. Chem. 265:11185–11192

Clary, D.O. & Wolstenholme, D.R. (1983a) Nucleotide sequence of a segment of *Drosophila* mitochondrial DNA that contains the genes for cytochrome c oxidase subunits II and III and ATPase subunit 6. Nucleic Acids Res. 11:4211–4227

Clary, D.O. & Wolstenholme, D.R. (1983b) Genes for cytochrome c oxidase subunit I, urf2, and 3 tRNA's in *Drosophila* mitochondrial DNA. Nucleic Acids Res. 11:6859–6872

Coruzzi, G. & Tzagoloff, A. (1979) Assembly of mitochondrial membrane systems. DNA sequence of subunit 2 of yeast cytochrome oxidase. J. Biol. Chem. 254:9324–9330

Dayhoff, M.O., Schwartz, R.M. & Orcutt, B.C. (1978) A model of evolutionary change in proteins. In: Dayhoff, M.O. (ed) Atlas of protein sequence and structure, vol. 5, suppl. 3. National Biomedical Research Foundation, Washington, D.C., pp. 345–352

Degli Esposti, M., Ghelli, A., Luchetti, R., Crimi, M. & Lenaz, G. (1989) New approaches to the prediction of the folding of membrane proteins with redox function. Ital. J. Biochem. 38:1–22

Devereux, J., Haeberli, P. & Smithies, O. (1984) A comprehensive set of sequence analysis programs for the VAX. Nucleic Acids Res 12:387–395

Doolittle, R.F. (1981) Similar amino acid sequences: chance or common ancestry? Science 214:149–159

Doolittle, R.F. (1986) Of urfs and orfs. A primer on how to analyze derived amino acid sequences. University Science Books, Mill Valley, California

Einarsdòttir, O., Killough, P.M., Fee, J.A. & Woodruff, W.H. (1989) An infrared study of the binding and photodissociation of carbon monoxide in cytochrome ba$_3$ from *Thermus thermophilus*. J. Biol. Chem. 264:2405–2408

Fee, J.A., Kuila, D., Mather, M.W. & Yoshida, T. (1986) Respiratory proteins from extremely thermophilic, aerobic bacteria. Biochim. Biophys. Acta 853:153–185

Fee, J.A., Mather, M.W., Springer, P., Hensel, S. & Buse, G. (1988) Isolation and partial sequence of the A-protein gene of *Thermus thermophilus* cytochrome c$_1$ aa$_3$. Ann. New York Acad. Sci. 550:33–38

Gelles, J., Blair, D.F. & Chan, S.I. (1986) The proton-pumping site of cytochrome c oxidase: a model of its structure and mechanism. Biochim. Biophys. Acta 853:205–236

Gribskov, M., McLachlan, A.D. & Eisenberg, D. (1987) Profile analysis: detection of distantly related proteins. Proc. Natl. Acad. Sci. USA 84:4355–4358

Hata, A., Kirino, Y., Matsuura, K., Itoh, S., Hiyama, T., Konishi, K., Kita, K. & Anraku, Y. (1985) Assignment of ESR signals of *Escherichia coli* terminal oxidase complexes. Biochim. Biophys. Acta 810:62–72

Holm, L., Saraste, M. & Wikström, M. (1987) Structural models of the redox centres in cytochrome oxidase. EMBO J. 6:2819–2823

Hon-nami, K. & Oshima, T. (1984) Purification and characterization of cytochrome c oxidase from *Thermus thermophilus* HB8. Biochemistry 23:454–460

Ishizuka, M., Machida, K., Shimada, I. & Sone, N. (1990) (submitted)

Kita, K., Konishi, K. & Anraku, Y. (1984) Terminal oxidases of *Escherichia coli* aerobic respiratory chain. J. Biol. Chem. 259:3368–3374

Kyte, J. & Doolittle, R.F. (1982) A simple method for displaying the hydropathic character of a protein. J. Mol. Biol. 157:105-132

Martin, C.T., Scholes, C.P. & Chan, S.I. (1988) On the nature of cysteine coordination to $Cu_A$ in cytochrome c oxidase. J. Biol. Chem 263:8420-8429

Mather, M.W. (1988) Base composition-independent hybridization in dried agarose gels: screening and recovery for cloning of genomic DNA fragments. Bio Techniques 6:444-447

Mather, M.W., Springer, P., & Fee, J.A. (1990) (submitted)

Michel, H, Weyer, K.A., Gruenberg, H., Dunger, I., Oesterhelt, D. & Lottspeich, F. (1986) The 'light' and 'medium' subunits of the photosynthetic reaction centre from *Rhodospeudomonas viridis*: isolation of the genes, nucleotide and amino acid sequence. EMBO J. 5:1149-1158

Millet, F., de Jong, C., Paulson, L. & Capaldi, R.A. (1983) Identification of specific carboxylate groups on cytochrome c oxidase that are involved in binding cytochrome c. Biochemistry 22:546-552

Müller, M., Schläpfer, B. & Azzi, A. (1988a) Preparation of a one-subunit cytochrome oxidase from *Paracoccus denitrificans*: spectral analysis and enzymatic activity. Biochemistry 27:7546-7551

Müller, M., Schläpfer, B. & Azzi, A. (1988b) Cytochrome c oxidase from *Paracoccus denitrificans*: both hemes are located in subunit I. Proc. Natl. Acad. Sci. USA 85:6647-6651

Palmer, G. (1987) Cytochrome oxidase: a perspective. Pure Appl. Chem. 59:749-758

Puustinen, A., Finel, M. Virkki, M. & Wikström, M. (1989) Cytochrome o (bo) is a proton pump in *Paracoccus denitrificans* and *Escherichia coli*. FEBS Lett. 249:163-167

Raitio, M. Tuulikki, J. & Saraste, M. (1987) Isolation and analysis of the genes for cytochrome c oxidase in *Paracoccus denitrificans*. EMBO J. 6:2825-2833

Robillard, G.T. & Lolkema, J.S. (1988) Enzymes II of the phosphoeno/pyruvate-dependent sugar transport systems: a review of their structure and mechanism of sugar transport. Biochim. Biophys. Acta 947:493-519

Salerno, J.C., Bolgiano, B. & Ingledew, W.J. (1989) Potentiometric titration of cytochrome-bo type quinol oxidase of *Escherichia coli*: evidence for heme-heme and copper-heme interaction. FEBS Lett. 247:101-105

Sanger, F., Nicklen, S. & Coulson, A.R. (1977) DNA sequencing with chain terminating inhibitors. Proc. Natl. Acad. USA 74:5463-5467

Saraste, M., Raitio, M., Tuulikki, J. & Perämaa, A. (1986) A gene in *Paracoccus* for subunit III of cytochrome oxidase. FEBS Lett. 206:154-156

Sone, N., Yanagita, Y., Hon-Nami, K., Fukumori, Y. & Yamanaka, T. (1983) Proton-pump activity of *Nitrobacter agilis* and *Thermus thermophilus* cytochrome c oxidases. FEBS Lett 155:150-154

Sone, N., Yokoi, F., Fu, T., Ohta, S. Metso, T., Raitio, M. & Saraste, M. (1988) Nucleotide sequence of the gene coding for cytochrome oxidase subunit I from the thermophilic bacterium PS3. J. Biochem. (Tokyo) 103:606-610

Steffens, G.J. & Buse, G. (1979) Studies on cytochrome c oxidase, IV. Primary structure and function of subunit II. Hoppe-Seyler's Z. Physiol. Chem. 360:613-619

Steinrücke, P., Steffens, G.C.M., Panskus, G., Buse, G. & Ludwig, B. (1987) Subunit II of cytochrome c oxidase from *Paracoccus denitrificans*. DNA sequence, gene expression and the protein. Eur. J. Biochem. 167:431-439

Sweet, R.M. & Eisenberg, D. (1983) Correlation of sequence hydrophobicities measures similarity in three-dimensional protein structures. J. Mol. Biol. 171:479-488

Thalenfeld, B.E. & Tzagoloff, A. (1980) Assembly of the mitochondrial membrane system. Sequence of the Oxi2 gene of yeast mitochondrial DNA. J. Biol. Chem. 255:6173-6180

Wikström, M., Saraste, M. & Pentillä, T. (1985) Relationships between structure and function in cytochrome oxidase. In: Martonosi, A.M. (ed) The enzymes of biological membranes, vol. 4. Plenum, New York

Woese, C.R. (1987) Bacterial evolution. Microbiol. Rev. 51:221-271

Yoshida, T. & Fee, J.A. (1984) Studies on cytochrome c oxidase activity of the cytochrome $c_1aa_3$ complex from *Thermus thermophilus*. J. Biol. Chem. 529:1031-1036

Yoshida, T., Lorence, R.M., Choc, M.G., Tarr, G.E., Findling, K.L. & Fee, J.A. (1984) Respiratory proteins from the extremely thermophilic aerobic bacterium, *Thermus thermophilus*. J. Biol. Chem. 259:112-123

Zimmermann, B.H., Nitsche, C.I., Fee, J.A., Rusnak, F. & Münck, E. (1988) Properties of a copper-containing cytochrome $ba_3$: a second terminal oxidase from the extreme thermophile *Thermus thermophilus*. Proc. Natl. Acad. Sci. USA 85:5779-5783

# The Problem of the Electrochemical Proton Potential in Alkaliphilic Bacteria

D.M. IVEY, D.B. HICKS, A.A. GUFFANTI, G. SOBEL, and T.A. KRULWICH[1]

## 1 Introduction

This discussion will focus on the aerobic, nonmarine, alkaliphilic *Bacillus* species that exhibit optimal growth at pH values in the range of pH 10–11. Such bacteria are readily isolated from natural and industrial enrichments. They have also been isolated from conventional soils that may have suitable microenvironments for their growth (reviewed in Grant and Tindall 1986; Krulwich and Guffanti 1989a). These organisms are of considerable applied interest largely because the substantial number of extracellular enzymes produced by various alkaliphilic *Bacillus* species often exhibit unusual alkalistability (Horikoshi 1988). Our own investigative interest arises from several fundamental bioenergetic problems that confront an obligately aerobic bacterium that grows at very high pH. Specifically, the alkaliphilic *Bacillus* species must, and indeed do, maintain a cytoplasmic pH that is much more acidic than the external medium. During active growth or respiration in the presence of a substrate, alkaliphilic *Bacillus firmus* RAB and related strain OF4, for example, maintain a cytoplasmic pH of 8.2–8.5 during growth at pH 10.5 (Krulwich and Guffanti 1989a; and see Table 1 below). The alkaliphilic *Bacillus* species are thus far all found to contain extremely high contents of membrane-associated respiratory chain components, with complexes that resemble those of other bacteria and effectively pump protons and only protons (Krulwich and Guffanti 1989a,b), in contrast to at least some marine bacteria that pump both $Na^+$ and protons during respiration (Unemoto et al. 1990). The net acidification of the cytoplasm is achieved through the concomitant function of the respiratory chain and an electrogenic $Na^+/H^+$ antiporter (Krulwich and Guffanti 1989b).

The maintenance of a cytoplasm that is over two pH units lower than the pH of the external milieu provides a remarkable biological example of pH homeostasis, but also presents a further bioenergetic problem. The latter arises from the need of an aerobe to carry out bioenergetic work that, in the widely accepted chemiosmotic model (Mitchell 1961), depends upon an electrochemical gradient of protons ($\Delta p$), outside acid and positive, across the coupling membrane. Since the pH gradient ($\Delta pH$) component of the alkaliphiles is reversed in orientation and substantial in magnitude under optimal growth conditions, the total bulk $\Delta p$ is low, in spite of a transmembrane electrical potential $\Delta\psi$ that is conventional in both orientation and magnitude (Table 1). For many bioenergetic functions, this

[1] Department of Biochemistry, Mount Sinai School of Medicine of CUNY, New York, New York, USA

41. Colloquium Mosbach 1990
The Molecular Basis of Bacterial Metabolism

**Table 1.** The magnitudes of the electrochemical proton gradients ($\Delta p$) and phosphorylation potentials ($\Delta G_p$) in cells of *B. firmus* OF4 growing at either pH 7.5 or 10.5[a]

| | Cells grown at pH 7.5 | Cells grown at pH 10.5 |
|---|---|---|
| $\Delta\psi$, mV | -139 | -182 |
| $\Delta$pH, units | 0 | 2.2 |
| $\Delta$pH, mV | 0 | +131 |
| $\Delta p$, mV | -139 | - 51 |
| $\Delta G_p$, kcal/mol | 10.7 | 10.5 |
| $\Delta G_p$, mV | -461 | -452 |
| $\Delta G_p/\Delta p$ | 3.3 | 8.8 |

[a]The cells were grown to mid-logarithmic phase at pH 7.5 or 10.5 in the D, L-malate-containing medium described previously (Guffanti et al. 1986). The $\Delta\psi$ and $\Delta$pH were measured directly in suspensions of the growing cells from the distributions, respectively, of 4 μM[$^3$H]-tetraphenylphosphonium (Schuldiner and Kaback 1975) and [$^{14}$C] 5,5-dimethyl-2,4-oxazolidine or 2 μM [$^{14}$C] methylamine (Zilberstein et al. 1982); the assays were carried out by filtration. ATP ADP, and inorganic phosphate were measured in the same cell samples as described previously (Guffanti et al. 1984) for calculation of the $\Delta G_p$.

problem of a low $\Delta p$ is bypassed by the organisms' use of an electrochemical gradient of $Na^+$ instead of an electrochemical gradient of protons (Krulwich and Guffanti 1989a,b). This solution is made possible by the $Na^+/H^+$ antiporter, which not only functions in pH homeostasis, but also converts the initial electrochemical proton gradient that is formed from respiration into a smaller $\Delta p$ and a larger electrochemical sodium gradient.

Importantly, however, the central bioenergetic work process for an aerobe, oxidative phosphorylation, is apparently completely coupled to protons in the extreme, nonmarine alkaliphiles. Thus a dilemma exists: the bulk $\Delta p$ is insufficient to account for the phosphorylation potentials ($\Delta G_p$) observed, if synthesis of ATP occurs by a chemiosmotic mechanism employing a $H^+$/ATP stoichiometry (taken at the $\Delta G_p/\Delta p$ ratio in the chemiosmotic formulation) that is invariant and near the generally observed value of 3 (Senior 1988).

## 2 Results and Discussion

### *2.1 The Quantitative and Qualitative Problem with Respect to Oxidative Phosphorylation and the Low $\Delta p$ of the Alkaliphiles*

The data shown in Table 1 for the facultative alkaliphile *B. firmus* OF4 are values for the $\Delta p$ components and the $\Delta G_p$ of cells that were actively growing at either pH 7.5 or pH 10.5. Actively growing alkaliphile cells, or cells that are given substrate in suspension, retain cytoplasmic pH values below 8.5, whereas our earliest measurements of the cytoplasmic pH of *Bacillus alcalophilus* and *B. firmus* RAB had indicated higher cytoplasmic pH values (closer to 9 at an external pH of 10.5) because they were conducted on cells that had been suspended in buffer without substrate. Although, as shown in Table 1, alkaliphiles generally exhibit higher $\Delta\psi$

values at very alkaline pH than at near neutral pH, the total bulk $\Delta p$ is much lower at near neutral pH because of the "reversed" $\Delta pH$, whereas the phosphorylation potentials at the two growth pH values are comparable. Calculation of the $\Delta G_p/\Delta p$ ratios, then, yields a rather conventional value of 3.3 for cells growing at pH 7.5, and a much higher value of 8.8 for cells growing at pH 10.5 (Table 1).

Conceivably, the ATPase that catalyzes synthesis might utilize close to nine protons per ATP synthesized at very alkaline pH values, while functioning at a $H^+$/ATP ratio of approximately 3 at lower pH. This is apparently not, however, the resolution of the quantitative dilemma; nor is the dilemma resolved by the possibility that, in spite of the use of several methods and various calibrations, the $\Delta\psi$ values are grossly underestimated for some unknown technical reason at high pH only. Were either of those resolutions correct, artificially imposed diffusion potentials of known magnitude would be expected to function identically at the two pH values. Precisely the opposite result is consistently observed. A valinomycin-mediated potassium diffusion potential, calculated and confirmed by measurement to be in the range of -160 to -170 mV in magnitude, leads to rapid synthesis of ATP by starved cells or appropriately loaded membrane vesicles of the alkaliphile at neutral pH (again yielding conventional $\Delta G_p/\Delta p$ ratios), but fails to result in any synthesis at pH values above 9–9.2 (Guffanti et al. 1984; Krulwich and Guffanti 1989b). Had either a variable $H^+$/ATP stoichiometry or an underestimated $\Delta\psi$ from respiration at high pH been the cause of the discrepancy between the putative driving force and the ATP synthesized, comparable diffusion potentials should have resulted in comparable synthesis at the two pH values. Additional experiments show (1) that the diffusion potentials at high pH are completely competent to drive $Na^+$ coupled solute uptake; (2) the simultaneous imposition of a large $Na^+$ chemical gradient does not result in ATP synthesis at the alkaline pH (as it would if a $Na^+$-coupled synthase operated at high pH); and (3) the rate of ATP hydrolysis is no greater at alkaline than at neutral pH (Guffanti et al. 1985; Krulwich and Guffanti 1989b). Finally, the possibility that the ATPase is sequestered in some intracytoplasmic membrane compartment has been eliminated by immunoelectron microscopic studies (Rohde et al. 1989).

## 2.2 *The Properties of the $F_1F_0$-ATPase Alkaliphilic B. firmus and Evidence that It is Exclusively Proton-Coupled*

As indicated above, imposition of large chemical gradients of $Na^+$, even in the presence of a diffusion potential, fails to result in ATP synthesis by starved alkaliphile cells at high pH. This is true under a variety of experimental conditions, including conditions in which imposition of a chemical gradient of protons does lead to ATP synthesis (Guffanti and Krulwich 1988). Nor is ATP synthesis by cells or appropriate vesicles either dependent upon or stimulated by added $Na^+$ (Krulwich and Guffanti 1989a). These observations essentially rule out the possible involvement of a $Na^+$-coupled synthase in the nonmarine alkaliphile. Studies of the ATPase activity in membranes and in purified and reconstituted preparations further support this conclusion.

The total membrane-associated ATPase activity (hydrolytic) or *B. firmus* RAB is inhibited by at least 80% upon treatment of the membranes with antibody that was raised against purified $F_1$-ATPase. The $F_1$-ATPase has a conventional 5-subunit structure. Neither that purified $F_1$-ATPase nor the membranes exhibit $Na^+$-stimulated ATPase activity (Hicks and Krulwich 1986). More recently, crude octylglucoside extracts of membranes from the related alkaliphilic *B. firmus* OF4 have been reconstituted into proteoliposomes, and purified $F_1F_0$-ATPase has similarly been reconstituted. Like the alkaliphile $F_1$, the holoenzyme appears to have a conventional subunit structure with respect to the number of polypeptides and their estimated molecular sizes. Although proteoliposomes reconstituted with crude octylglucoside extracts from alkaliphile membranes catalyze secondary $Na^+/H^+$ antiport, no primary, ATP-dependent $^{22}Na^+$ uptake can be demonstrated in these or the more purified preparations using a variety of conditions, including those used to demonstrate ATP-dependent $Na^+$ pumping by the ATPase from *Propionigenium modestum* (Laubinger and Dimroth 1988). The $F_1F_0$-ATPase of the alkaliphile does exhibit proton pumping, as assayed by a fluorescence quenching assay, as well as uncoupler (carbonyl cyanide *m*-chlorophenylhydrazone, CCCP)-stimulated ATPase (hydrolytic) activity, and energy-dependent ATP-$^{32}$Pi exchange. The absence of $Na^+$ stimulation of the two latter activities is shown in Figs. 1 and 2. Moreover, ATP-$^{32}$Pi exchange is completely sensitive to CCCP, as indicated in the legend to Fig. 2.

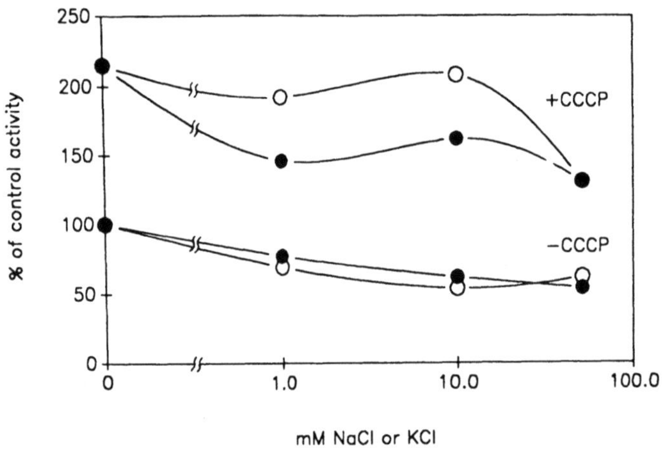


**Fig. 1.** Effect of NaCl or KCl on coupled and uncoupled ATP hydrolysis in *B. firmus* OF4 $F_1F_0$-ATPase proteoliposomes. $F_1F_0$-ATPase was purified and reconstituted into proteoliposomes by octylglucoside dilution by the method of Racker et al. (1979) using soybean asolectin. The proteoliposomes were preincubated for 15 min at 37°C in a reaction mixture containing 20 mM Tricine-KOH, pH 8, 5 mM $MgCl_2$, the indicated concentrations of NaCl (*open circles*) or KCl (*closed circles*), and, for uncoupled ATPase activity, 10 μM CCCP. Reactions were initiated with 5 mM ATP (potassium salt) and terminated at 15 min by the addition of TCA. Following centrifugation, the supernatant fluids were analyzed for inorganic phosphate (Pi) content by the method of Le Bel et al. (1978). 100% activity was 0.65 μmol Pi $min^{-1}$ mg $protein^{-1}$

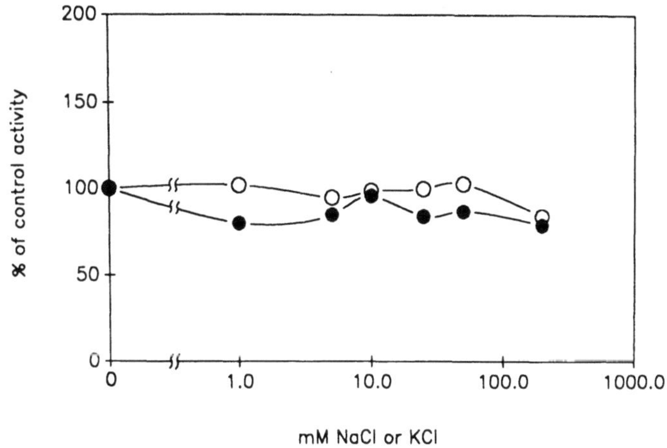


**Fig. 2.** Effect of NaCl or KCl on ATP-$^{32}$Pi exchange activity in *B. firmus* OF4 $F_1F_0$-ATPase proteoliposomes. Proteoliposomes containing the $F_1F_0$-ATPase were prepared as described in the legend to Fig. 1. They were preincubated for 10 min at 37°C in a reaction mixture containing 50 mM Tricine-KOH, pH 8, 5 mM $MgCL_2$, 3 mM potassium phosphate, pH 8 (containing $3 \times 10^5$ cpm $^{32}$Pi/$\mu$mol) and the indicated concentrations of NaCl (*open circles*) or KCl (*closed circles*). The exchange reaction was initiated with 5 mM ATP (potassium salt) and terminated at 30 min by the addition of TCA. Unreacted Pi was extracted from the deproteinized supernatant fluids by the method of Avron (1960). $AT^{32}P$ was estimated by liquid scintillation counting of aliquots of the aqueous phase. TCA was added just prior to ATP in the zero time controls. 100% activity was 18.3 nmol $AT^{32}P$ $min^{-1}$ mg $protein^{-1}$; 10 $\mu$M CCCP completely abolished activity

### *2.3 A Parallel Coupling Model that Accommodates the Bioenergetic Data and Encompasses Properties of the Alkaliphile Membrane*

We have proposed a parallel coupling model for oxidative phosphorylation by alkaliphiles (Krulwich and Guffanti 1989a,b). At near neutral pH, energy coupling for oxidative phosphorylation could occur by a chemiosmotic mechanism utilizing the substantial $\Delta p$, and yielding an apparently conventional relationship between that driving force and the observed $\Delta G_p$. At very alkaline pH, by contrast, we propose that the alkaliphile $F_1F_0$-ATPase restricts proton movement to and from the bulk phase by a gating mechanism. Control of the gate could be directly pH regulated, as has been proposed in another system (Horner and Moundrianakis 1986). Such a gating would explain why a diffusion potential that is competent for other bioenergetic work at high pH and competent for ATP synthesis at lower pH, fails to energize ATP synthesis at the high pH that is the organism's preferred growth pH. Coupling under these conditions occurs, then, not through the bulk $\Delta p$, be it a large imposed one or the low natural one, but rather through direct hand-off of protons from respiratory chain pumps to proton-accepting residues on the $F_0$ sector of the ATPase within the membrane. This process is proposed to be facilitated by the very high concentration of respiratory chain components that is observed in all the extreme, aerobic alkaliphiles and by a membrane lipid composition that may help to maximize the diffusion rate of those components and hence, their likely collision with the $F_1F_0$-ATPase. For this direct coupling mode, the restriction of proton flow to the bulk would prevent the protons that arrive at the ATPase from exiting into the bulk, rather than moving inward and catalyzing synthesis. The $\Delta p$, albeit low, is probably also essential for this reason; the extreme alkaliphiles do not exhibit ATP synthesis when this low $\Delta p$ is completely abolished. The direct coupling mode, as outlined, would be dependent upon specific membrane lipid properties that favor the diffusion of the relevant interacting species and their productive collision. The specific lipid species, their microviscosity, and their packing properties, could become an important feature of energy coupling. This is an important distinction between a direct proton pathway of the type discussed here and the chemiosmotic pathway through the bulk, in which the exclusive role of the membrane lipids is in providing a relatively ion impermeable barrier.

If the proton pathway during oxidative phosphorylation at pH 7.5 is entirely or at least largely through the bulk, as suggested by the conventional $\Delta G_p/\Delta p$ ratio, whereas the pathway at pH 10.5 is almost exclusively direct, then it should be possible to select mutants that have lost the direct coupling capacity. If the model outlined above is correct, then such mutants might involve decreased levels of proton pumps or altered membrane lipid properties; additional candidates may arise from the molecular biological studies described in the next section. Such mutants would be expected to have lost the ability to grow at very alkaline pH, while retaining the $Na^+$ cycle that achieves pH regulation. They would retain the ability to grow at pH 7.5. Using newly developed methods for achieving transposition in alkaliphilic *B. firmus* OF4, we are initiating the search for such pH-conditional phenotypes.

### 2.4 Molecular Biological Studies of the Genes Encoding the $F_1F_0$-ATPase from Alkaliphilic *B. firmus*

A parallel coupling model of the type outlined above does not necessarily involve any special properties of the ATPase itself, although a high concentration of proton-accepting residues within the membrane, e.g., a high *c*-subunit content per holoenzyme or even an unusual $F_0$ subunit structure might have evolved in this very extreme organism to make the localized mode particularly effective. With these possibilities in mind, genes encoding the enzyme have been cloned from *B. firmus* RAB and OF4 using probes prepared from genes encoding $F_1$ subunits in *Bacillus megaterium* (Brusilow et al. 1989). As shown in Fig. 3, the deduced sequence of the $\beta$ subunit of *B. firmus* RAB is extremely homologous to that from *B. megaterium*. These two *Bacillus* subunits are, in turn, most homologous with their counterpart from another *Bacillus* species, PS3, but also show strong homology with the $\beta$ subunits from less related species. This is as expected for the catalytic, hydrophilic polypeptide subunit. Moreover, the gene order of the $F_1$ genes of *B. firmus* RAB is the same as in other bacteria, and the gene encoding the *b* subunit of the $F_0$ is in the usual position, contiguous with the genes encoding the five $F_1$ subunits. The genes that encode the *a* and *c* subunits of the $F_0$ would be expected to be immediately upstream of the *b* gene. These most interesting genes and any regulatory elements associated with them are not yet identified and characterized in the alkaliphile.

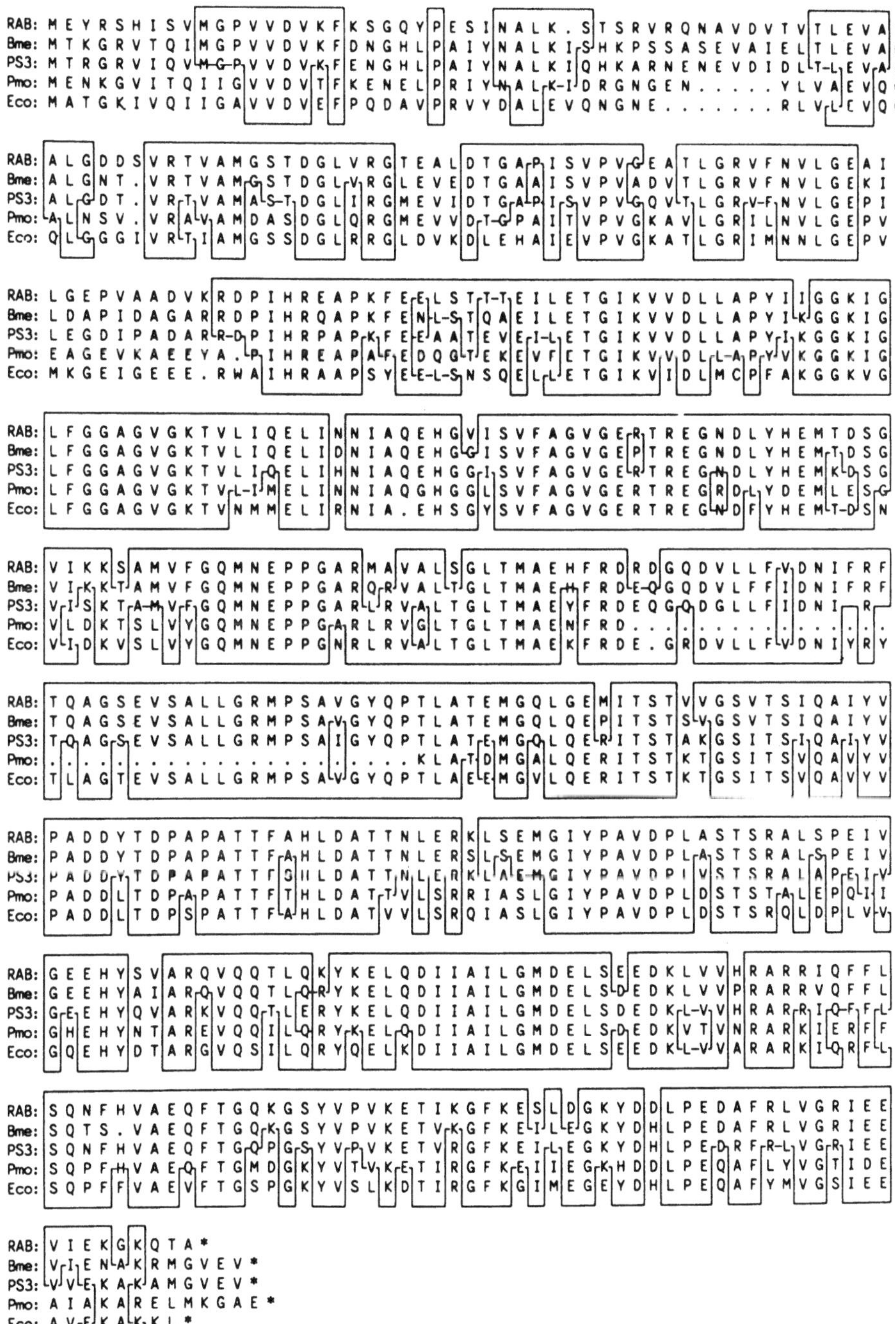

**Fig. 3.** Amino acid sequence of the $\beta$ subunit of the $F_1F_0$-ATPase from alkaliphilic *B. firmus* RAB, deduced from the gene sequence and shown in comparison with data from homologous subunits from other bacteria. The data shown are for: *RAB B. firmus* RAB (Ivey and Krulwich 1990); Bme *Bacillus megaterium* (Hawthorne and Brusilow 1988); *PS3* thermophilic *Bacillus* PS3 (Kagawa et al. 1986); *Pmo Propionigenium modestum* (Amann et al. 1988); *Eco Escherichia coli* (Walker et al. 1984). Conserved residues are indicated by the *enclosures*

*Acknowledgment.* This work was supported by research grant GM28454 from the National Institute of General Medical Sciences.

## References


Amann, R., Ludwig, W., Laubinger, W., Dimroth, P. & Schleifer, K.H. (1988) Cloning and sequencing of the genes encoding the beta subunit of the sodium ion translocating ATP synthase of *Propionigenium modestum*. FEMS Microbiol. Lett. 56:253–260

Avron, M. (1960) Photophosphorylation by swiss chard chloroplasts. Biochim. Biophys. Acta 40:257–272

Brusilow, W.S.A., Scarpetta, M.A., Hawthorne, C.A. & Clark, W.P. (1989) Organization and sequence of the genes coding for the proton-translocating ATPase of *Bacillus megaterium*. J. Biol. Chem. 264:1528–1533

Grant, W.D.& Tindall, B.J. (1986) The alkaline saline environment. In: Herbert, R.A., & Codd, G.A. (eds.) Microbes in extreme environments. Academic Press, Lond New York, pp. 25–54

Guffanti, A.A. & Krulwich, T.A. (1988) ATP synthesis is driven by an imposed ΔpH or $\Delta\tilde{\mu}_{H^+}$ but not by an imposed $\Delta pNa^+$ or $\Delta\tilde{\mu}_{Na^+}$ in alkalophilic *Bacillus firmus* OF4 at high pH. J. Biol. Chem. 263:14748–14752

Guffanti, A.A., Fuchs, R.T., Schneier, M., Chiu, E. & Krulwich, T.A. (1984) A transmembrane electrical potential generated by respiration is not equivalent to a diffusion potential of the same magnitude for ATP synthesis by *Bacillus firmus* RAB. J. Biol. Chem. 259:2971–2975

Guffanti, A.A., Chiu, E. & Krulwich, T.A. (1985) Failure of an alkalophilic bacterium to synthesize ATP in response to a valinomycin-induced potassium diffusion potential at high pH. Arch. Biochem. Biophys. 239:327–333

Guffanti, A.A., Finkelthal, O., Hicks, D.B., Falk, L., Sidhu, A., Garro, A. & Krulwich, T.A. (1986) Isolation and characterization of new facultatively alkalophilic strains of *Bacillus* species. J. Bacteriol. 167:766–773

Hawthorne, C.A. & Brusilow, W.S.A. (1988) Sequence of the genes for the $\beta$ and $\varepsilon$ subunits of the ATP synthase of *Bacillus megaterium* QMB1551. Biochem. Biophys. Res. Commun. 151:926–931

Hicks, D.B. & Krulwich, T.A. (1986) The membrane ATPase of alkalophilic *Bacillus firmus* RAB is an $F_1$-type ATPase. J. Biol. Chem. 261:12896–12902

Horikoshi, K. (1988) Genetic applications of alkalophilic microorganisms. In: Herbert, R.A. & Codd, G.A. (eds.) Microbes in extreme environments. Academic Press, Lond New York, pp. 297–315

Horner, R.D. & Moundrianakis, E.N. (1986) Effects of permeant buffers on the initiation of photosynchronous phosphorylation and postillumination phosphorylation in chloroplasts. J. Biol. Chem. 261:13408–13414

Ivey, D.M. & Krulwich, T.A. (1990) Sequence of the gene encoding the ATP synthase beta subunit from alkaliphilic *Bacillus firmus* RAB. Nucleic Acids Res. 18:1296

Kagawa, Y., Ishizuka, M., Saishu, T. & Nakao, S. (1986) Stable structure of thermophilic proton ATPase beta subunit. J. Biochem. 100:923–934

Krulwich, T.A. & Guffanti, A.A. (1989a) Alkalophilic bacteria. Ann. Rev. Microbiol. 43:435–463

Krulwich, T.A. & Guffanti, A.A. (1989b) The $Na^+$ cycle of extreme alkalophiles: a secondary $Na^+/H^+$ antiporter and $Na^+$/solute symporters. J. Bioenerg. Biomembr. 21:663–677

Laubinger, W. & Dimroth, P. (1988) Characterization of the ATP synthase of *Propionigenium modestum* as a primary sodium pump. Biochemistry 27:7531–7537

Le Bel, D., Poirer, G.D. & Beadoin, A.R. (1978) A convenient method for the ATPase assay. Anal. Biochem. 85:86–89

Mitchell, P. (1961) Coupling of phosphorylation to electron and hydrogen transfer by a chemiosmotic type of mechanism. Nature (Lond) 191:144–148

Racker, E., Violand, B., O'Neal, S., Alfonzo, M. & Telford, J. (1979) Reconstitution, a way of biochemical research: some new approaches to membrane-bound enzymes. Arch. Biochem. Biophys. 198:470–477

Rohde, M., Mayer, F., Hicks, D.B. & Krulwich, T.A. (1989) Immunoelectron microscopic localization of the $F_1F_0$-ATPase (ATP synthase) on the cytoplasmic membrane of alkalophilic *Bacillus firmus* RAB. Biochim. Biophys. Acta 985:233–235

Schuldiner, S. & Kaback, H.R. (1975) Membrane potentials and active transport in membrane vesicles from *E. coli*. Biochemistry 14:5451–5461

Senior, A.E. (1988) ATP synthesis by oxidative phosphorylation. Physiol. Rev. 68:177–230

Unemoto, T., Tokuda, H. & Hayashi, M. (1990) Primary sodium pumps and their significance in bacterial energetics. In: Ornston, Sokatch, J., Krulwich, T.A. (eds.) The bacteria, Vol. 12. Academic Press, Orlando, pp. 33–54

Walker, J.E., Saraste, M. & Gay, N.H. (1984) The *unc* operon: nucleotide sequence, regulation and structure of ATP synthase. Biochim. Biophys. Acta 768:164–200

Zilberstein, D., Agmon, V., Schuldiner, S. & Padan, E. (1982) The sodium/proton antiporter is part of the pH homeostasis mechanism in *Escherichia coli*. J. Biol. Chem. 257:3687–3691

# Energy Transductions by an Electrochemical Gradient of Sodium Ions

P. DIMROTH[1]

## 1 Introduction

All living cells have to maintain a membrane potential for vitality. This is established by electrogenic ion movements across the membrane, driven by either light or chemical energy, and provides an energy source for endergonic reactions associated with the membrane. The most common coupling ions of bacteria are protons, but in some cases, $Na^+$ ions play a central role in energy transducing reactions (Dimroth 1987). Light-energy conversion into an electrochemical gradient of protons ($\Delta\tilde{\mu}H^+$) is performed by phototrophic bacteria. Aerobic bacteria use the free energy of substrate oxidation with $O_2$ as electron acceptor for $H^+$ pumping (or $Na^+$ pumping by some marine organisms) out of the cells. Many anaerobic bacteria generate $\Delta\tilde{\mu}H^+$ by electron transport reactions with electron acceptors other than oxygen (e.g., nitrate, fumarate, sulfate, carbonate). In other anaerobes, the membrane potential has to be created by the $F_1F_0$ATPase operating in reverse, i.e., in pumping protons out of the cell upon ATP hydrolysis. Other anaerobes have developed a different mechanism for the generation of a membrane potential: they perform $Na^+$ pumping to the outside of the membrane upon decarboxylation of certain "energy-rich" carboxylic acids, e.g., oxaloacetate (Dimroth 1987, 1989). The catalysis by exchange carriers of an obligate exchange of differently charged ions is another alternative for membrane potential generation. An example for such a system is the oxalate/formate exchanger of *Oxalobacter formigenes* (Anantharam et al. 1989). An electrochemical gradient of $H^+$ or $Na^+$ generated by either of these reactions may be used as an energy source for solute uptake, ATP synthesis, or flagellar motion.

In this chapter, I want to focus on the structure and on the catalytic mechanism of the sodium transport decarboxylases. I will also present evidence for the physiological significance of the $Na^+$ ion gradients established by these enzymes for the bioenergetics of the respective bacteria, and in this context I review recent advances on structure and mechanism of the $Na^+$ translocating ATPase of *Propionigenium modestum*.

[1]Mikrobiologisches Institut der Eidgenössischen Technischen Hochschule. ETH-Zentrum, Schmelzbergstr. 7, 8092 Zürich, Switzerland

41. Colloquium Mosbach 1990
The Molecular Basis of Bacterial Metabolism

## 2 Results and Discussion

### 2.1 The Sodium Ion Transport Decarboxylases

#### 2.1.1 Physiological Function

The fermentation pathways performed by a number of anaerobic bacteria involve the decarboxylation of an "energy-rich" carboxylic acid ($\Delta G^{o\prime} \approx -28$ kJ/mol) (Dimroth 1987, 1989). It has been shown that this energy is used by certain decarboxylases to pump sodium ions out of the cells. This energy conservation mechanism has been detected for the oxaloacetate decarboxylase of *Klebsiella pneumoniae* fermenting citrate. The individual reactions of the citrate fermentation pathway are shown in Fig. 1. The first step is the cleavage of citrate to acetate and oxaloacetate, catalyzed by citrate lyase. This is succeeded by the decarboxylation of oxaloacetate to pyruvate and $CO_2$. Pyruvate is subsequently split to acetyl-CoA and formate by pyruvate formate lyase. Acetyl-CoA is converted to acetyl phosphate, from which ATP is formed by substrate-level phosphorylation. Thus, 1 mol of ATP is synthesized in these cells per mol of citrate fermented.

The oxaloacetate decarboxylase has some remarkable properties. It is firmly bound to the membrane, specifically activated by $Na^+$ ions, and functions as an electrogenic $Na^+$ pump, transporting 2 mol $Na^+$ ions out of the cell per mol oxaloacetate decarboxylated (Dimroth 1982a, 1982b, 1987). Also noteworthy is the presence of covalently bound biotin in this enzyme, the significance of which for the catalytic mechanism will be discussed in Section 2.1.3.

It is obvious that in generating a $Na^+$ ion gradient on oxaloacetate decarboxylation, *K. pneumoniae* conserves energy. This energy is required to drive the uptake of the growth substrate citrate into the cells. Citrate occurs at neutral pH mainly as the trianion, and to not be repelled by the membrane potential (inside negative), its uptake requires a cotransport with cations containing at least three positive charges. In studies with whole cells and with a reconstituted proteoliposomal system, the citrate carrier of citrate fermenting *K. pneumoniae* cells was shown to be an electroneutral $Na^+$-citrate symporter (Dimroth and Thomer 1986, 1990). The $Na^+$ cycle would be completed if two $Na^+$ ions were exported again by oxaloacetate decarboxylase, and the third $Na^+$ ion would leave the cell by the

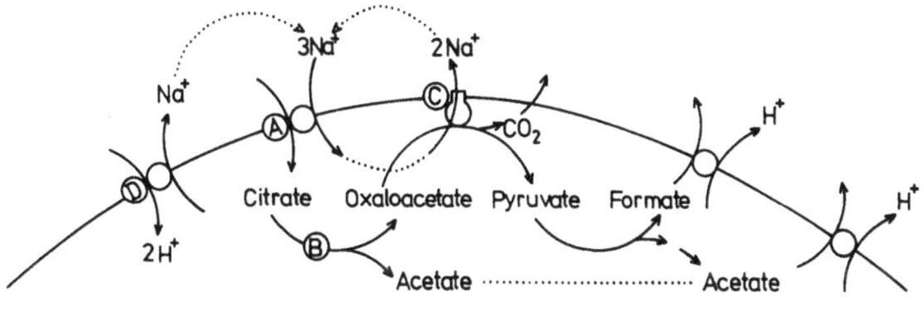


**Fig. 1.** Pathway of citrate fermentation in *Klebsiella pneumoniae* with a $Na^+$ circuit as a possible coupling mechanism between citrate uptake, oxaloacetate decarboxylation, and the $Na^+/H^+$ antiport system. *A* Citrate uptake system; *B* citrate lyase; *C* oxaloacetate decarboxylase; *D* $Na^+/H^+$ antiporter

$Na^+/H^+$ antiporter. Driving force for the antiporter could be obtained by $\Delta\psi$ created by oxaloacetate decarboxylase, and by a proton gradient generated by the extrusion of the acidic end products acetate, formate, or carbonate in symport with protons (Fig. 1). The $Na^+$-citrate symporter appears to be specifically designed for citrate-fermenting *K. pneumoniae* cells; under aerobic conditions the same cells express a $Na^+$-independent citrate carrier which probably uses $H^+$ as coupling ions (Dimroth and Thomer 1986); citrate uptake into other bacterial cells is a symport with divalent metal ions (Bergsma and Konings 1983).

A sodium ion-pumping oxaloacetate decarboxylase is also present in citrate fermenting *Salmonella typhimurium* cells (Wifling and Dimroth 1989), where the physiological role of the $Na^+$ gradient established by this enzyme may be the same as in *K. pneumoniae*. Other enzymes belonging to this group are methylmalonyl-CoA decarboxylase (Hilpert and Dimroth 1983) and glutaconyl-CoA decarboxylase (Buckel and Semmler 1983). Methylmalonyl-CoA decarboxylase has been found in *Veillonella alcalescens* fermenting lactate and in *Propionigenium modestum* fermenting succinate. As *V. alcalescens* lacks a membrane-bound ATPase (unpubl. observations) and as in *P. modestum*, the methylmalonyl-CoA decarboxylase-derived $Na^+$ gradient is the only energy source for ATP synthesis (Hilpert et al. 1984), generation of a membrane potential in these bacteria depends entirely on the decarboxylase action. Glutaconyl-CoA decarboxylase occurs in a number of anaerobic bacteria which ferment glutamate (Buckel and Semmler 1983). The physiological role of the $Na^+$ gradient established by the decarboxylase in these bacteria is unknown.

### *2.1.2 Purification and Properties*

All sodium ion-transport decarboxylases studied share a number of common properties (Dimroth 1987, 1989): (1) they are tightly membrane-bound enzyme complexes; (2) they are specifically activated by $Na^+$ ions; (3) they catalyze an electrogenic $Na^+$ transport across the membrane to the side opposite to that of substrate decarboxylation; (4) they contain covalently bound biotin; (5) they have a similar subunit composition; and (6) analogous subunits have similar functions.

A number of these enzymes have been highly purified using very similar protocols (Dimroth 1986; Hilpert and Dimroth 1986; Buckel 1986). The bacteria were disrupted by passage through a French-pressure-chamber, the membranes isolated by fractionated centrifugation, and the enzymes extracted with Triton X-100. Each solubilized decarboxylase was passed over a monomeric avidin-Sepharose affinity column, from which it was eluted specifically with biotin. A comparison of the subunit composition of several sodium transport decarboxylases is shown in Table 1. The oxaloacetate decarboxylases of *K. pneumoniae* (Dimroth and Thomer 1983) and *S. typhimurium* (Wifling and Dimroth 1989) have an almost identical pattern of three subunits ($\alpha$, $\beta$, $\gamma$). Subunits corresponding in molecular weight to the $\alpha$-, $\beta$-, and $\gamma$-chains are also present in methylmalonyl-CoA decarboxylase (Hilpert and Dimroth 1983) and subunits corresponding to the $\alpha$- and $\beta$-chains are present in glutaconyl-CoA decarboxylase (Buckel and Semmler 1983; Buckel and Liedtke 1986). A characteristic of these two enzymes is the presence of a separate biotin carrier protein subunit, whereas the biotin is bound to the $\alpha$-chain

**Table 1.** Subunit composition of sodium transport decarboxylases

| Decarboxylase | Organism | Molecular mass (kDa) | | | |
|---|---|---|---|---|---|
| | | α-Chain | β-Chain | γ-Chain | Biotin-carrier protein |
| Oxaloacetate | *Klebsiella pneumoniae* | 64 | 35 | 9 | α-chain |
| | *Salmonella typhimurium* | 63 | 34 | 10 | α-chain |
| Methylmalonyl-CoA | *Veillonella alcalescens* | 60 | 33 | 14 | 18.5 |
| Glutaconyl-CoA | *Acidaminococcus fermentans* | 60 | 35 | – | 25 |
| | *Peptostreptococcus assaccharolyticus* | 60 | ? | | 15 |
| | *Clostridium symbiosum* | 60 | ? | | 17, 20 |

in oxaloacetate decarboxylase. It is remarkable that the sodium transport decarboxylases, which have three different substrate specificities and are found in bacteria that are not closely related phylogenetically, share a similar composition of the α- and β-subunits. It will be shown below that these subunits perform analogous functions within each enzyme complex. The sodium transport decarboxylases are therefore a closely related family of enzymes.

### 2.1.3 Catalytic Mechanism

The decarboxylation mechanism of all sodium transport decarboxylases involves two distinct steps, each of which is catalyzed by a corresponding subunit of these enzyme complexes (Dimroth 1987, 1989). The reaction cycle is initiated by a transfer of the carboxyl group from the substrate to the prosthetic biotin group on the enzyme [carboxyltransferase; Eq. (1)]

$$\text{R-COO}^- + \text{biotin-protein} \rightleftharpoons \text{RH} + {}^-\text{OOC-biotin-protein}. \quad (1)$$

This reaction is catalyzed by the α-subunit of each enzyme complex (Dimroth and Thomer 1983; Hoffmann et al. 1989; Buckel and Liedtke 1986). It is completely independent from the presence of $Na^+$ ions. All different α-subunits are peripheral membrane proteins that, when dissociated from the other subunits, are completely soluble in the absence of detergent. The carboxyltransferase reaction is succeeded by the $Na^+$-dependent decarboxylation of the carboxybiotin-protein [lyase; Eq. (2)], which is believed to be coupled to $Na^+$ translocation:

$$\begin{aligned} &{}^-\text{OOC-biotin-protein} + H^+ + 2\,Na^+_{in} \\ &\rightleftharpoons CO_2 + \text{biotin-protein} + 2\,Na^+_{out}. \end{aligned} \quad (2)$$

This reaction is catalyzed by the β- (or β + γ)-subunits which are firmly embedded within the lipid bilayer.

To elucidate the above-mentioned functions of the subunits of the sodium transport decarboxylases, the enzyme complexes had to be dissociated. Oxaloacetate decarboxylase was completely split into subunits upon freezing and thawing in the presence of chaotropic salts, e.g., $LiClO_4$ (Dimroth and Thomer 1988). The dissociated enzyme was fractionated by chromatography on avidin-Sepharose into the ($\beta + \gamma$)-subunits and the biotin-containing $\alpha$-subunit (Fig. 2). Each fraction alone was inactive in the oxaloacetate decarboxylation reaction, but after combining the two fractions, the decarboxylase activity was recovered. The $\alpha$-subunit as well as the $\beta$- and $\gamma$-subunits have therefore been isolated in a catalytically active state. For methylmalonyl-CoA decarboxylase (Hoffmann et al. 1989) and glutaconyl-CoA decarboxylase (Buckel and Liedtke 1986), thus far only the $\alpha$-subunit and the biotin-protein have been dissociated from the complex with retention of catalytic activity.

### 2.1.4 *Structural Studies*

A summary of the structure/function relationship of oxaloacetate decarboxylase is shown in Fig. 3. The peripheral $\alpha$-subunit consists of two different domains, the carboxyltransferase domain ($\alpha_1$) and the biotin-carrier domain ($\alpha_2$) which can be easily separated by limited proteolysis with, for example, trypsin (Dimroth and Thomer 1983). Investigations by electron microscopy revealed a cleft in the $\alpha$-subunit with the prosthetic biotin group located at its bottom in close proximity to the $\beta$- and $\gamma$-subunits (Däkena et al. 1988). This location is in accord with the catalytic mechanism, which predicts the flip-flop movement of the biotin between the carboxyltransferase site on the $\alpha$-subunit and the lyase catalytic site on the $\beta$- (or $\beta + \gamma$)-subunits (Dimroth 1982b; Dimroth and Thomer 1983). The complete

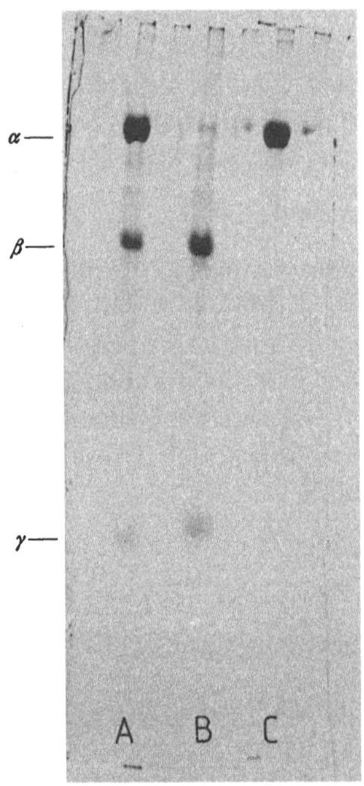


**Fig. 2.** Dissociation of oxaloacetate decarboxylase and fractionation by monomeric avidin-Sepharose chromatography. *A* Oxaloacetate decarboxylase; *B* fraction I, not retained by the column; *C* fraction II, retained and eluted with biotin from the column (Dimroth and Thomer 1988)

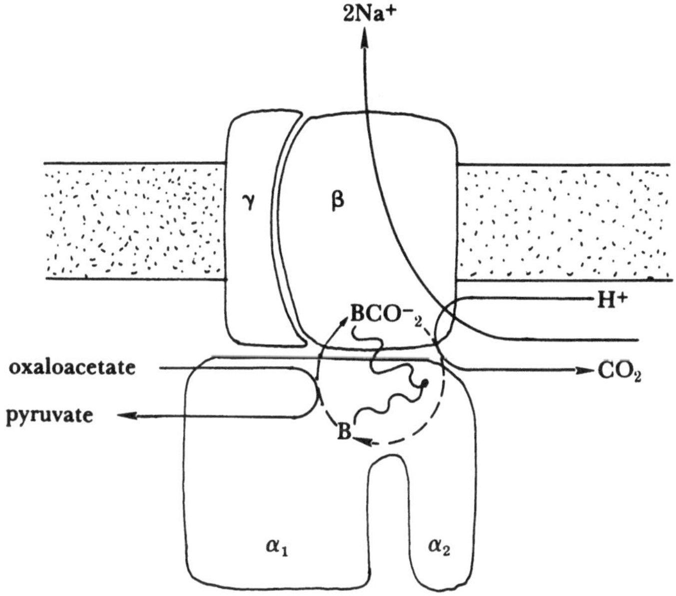


**Fig. 3.** *A* hypothetical model linking structure and function of the oxaloacetate decarboxylase subunits. *B* Biotin (Dimroth 1989)

primary structure of oxaloacetate decarboxylase has been determined by nucleotide sequencing of the cloned gene and by partial protein sequencing (Schwarz et al. 1988; Laußermair et al. 1989). The sequence of the N-terminal domain of the α-subunit ($\alpha_1$) is strikingly homologous to the 5S subunit of transcarboxylase that catalyzes exactly the same carboxyltransfer reaction, and the sequence of its C-terminal domain ($\alpha_2$) is homologous to that of other biotin-containing peptides, suggesting that these structures evolved from a common ancestor. A remarkable sequence that is unique to the α-chain of oxaloacetate decarboxylase is found between residues -59 and -28 upstream of the biotin-binding lysine residue. This sequence is: A-A-A-P-A-P-A-P-A-P-A-P-A-S-A-P-A-A-A-A-P-A. Such an extended area of mostly alanine and proline residues may provide a point of flexibility within the protein structure to allow the flip-flop movement of the biotin group between the two catalytic centers on the α- and β-subunit, respectively. It is interesting in this context that the dihydrolipoamide acetyltransferase subunit of the pyruvate dehydrogenase multienzyme complex of *Escherichia coli* contains a sequence rich in alanine and proline residues in a similar distance from the dihydrolipoamide prosthetic group (Stephens et al. 1983). This part of the protein is highly mobile, as shown by NMR spectroscopic techniques (Radford et al. 1987).

The sequence of the β-subunit contains extended stretches of hydrophobic amino acid residues, in accord with the location of this subunit firmly embedded within the membrane. Based on hydropathy and acrophilicity calculations, the secondary structure of the β-subunit could be formed with five or six intramembrane helical segments, the one with six helices being more likely (Laußermair et al. 1989). Hydropathy calculations defined one transmembrane helix in the

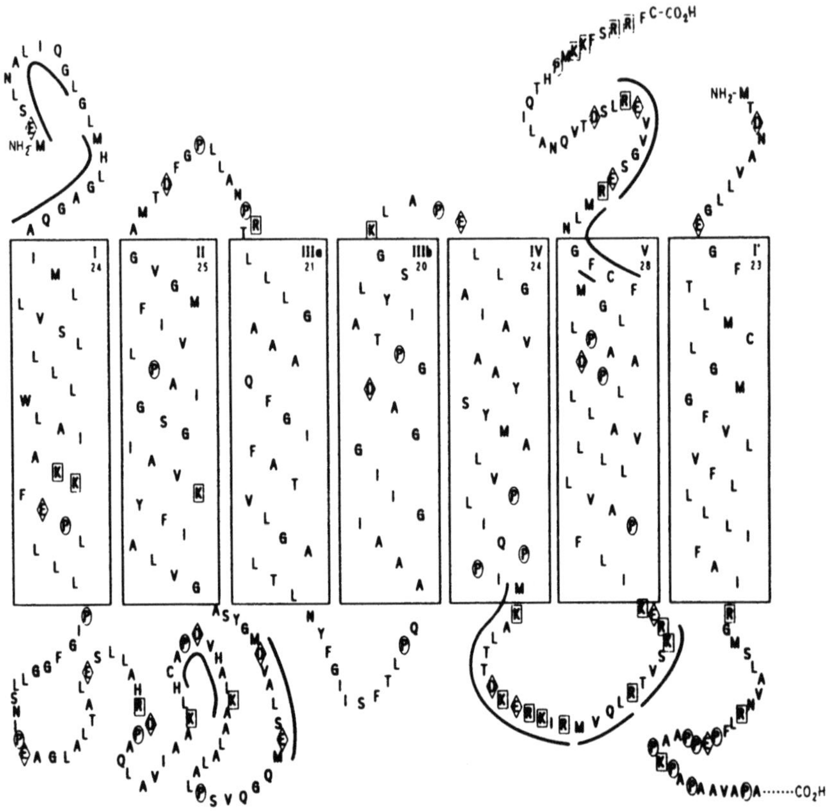


**Fig. 4.** Secondary structural model of the membrane-bound subunits $\beta$ and $\gamma$ of oxaloacetate decarboxylase with seven helices. The six helices on the *left* belong to the $\beta$- and the one on the *right* belongs to the $\gamma$-subunit. Amino-acid sequences obtained by automated Edman degradation are indicated with a *line*. Proline residues are *circled*, basic residues *boxed*, and acidic residues indicated by *diamonds* (Laußermair et al. 1989)

amino-terminal part of the $\gamma$-subunit and a hydrophilic carboxyl-terminal part that is certainly not embedded within the lipid bilayer (Laußermair et al. 1989). A secondary structural model indicating the arrangement of these helices in the membrane is shown in Fig. 4.

## 2.2 *The $Na^+$ Cycle in Propionigenium modestum*

*Propionigenium modestum* is the most prominent example of a bacterium depending on a $Na^+$ cycle for energy transduction. The strict anaerobe grows from the fermentation of succinate to propionate and $CO_2$ by the reactions summarized in Fig. 5 (Hilpert et al. 1984). The only step in the fermentation pathway that is sufficiently exergonic to be used for energy conservation is the decarboxylation of (*S*)-methylmalonyl-CoA ($\Delta G^{\circ\prime} \approx -27$ kJ/mol). The corresponding decarboxylase catalyzes the transport of two $Na^+$ ions per reaction out of the cell in an electrogenic manner (Dimroth and Hilpert 1984). The resulting $\Delta\tilde{\mu}Na^+$ serves *P. modestum* as

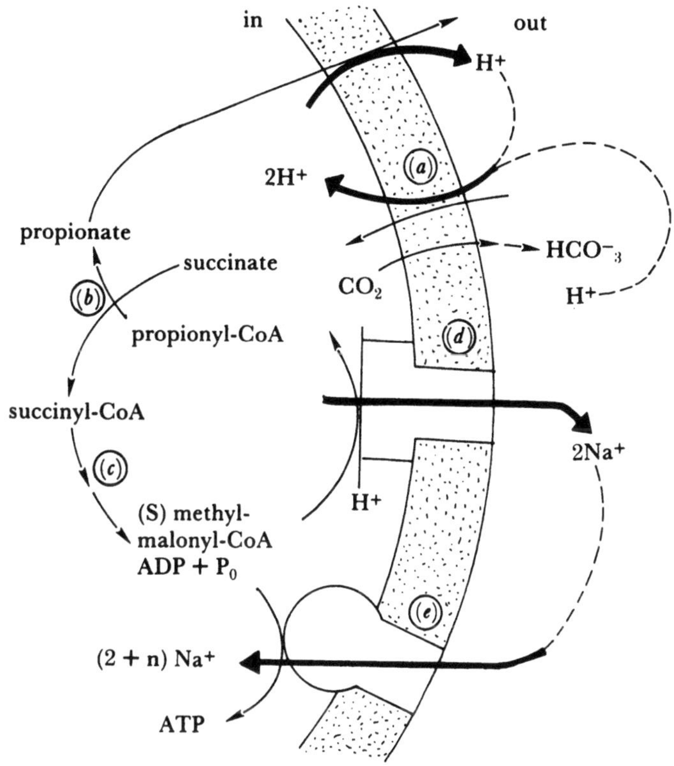


**Fig. 5.** Energy metabolism of *Propionigenium modestum* with a $Na^+$ cycle coupling the exergonic decarboxylation of (*S*)-methylmalonyl-CoA to endergonic ATP synthesis. A hypothetical proton circuit could couple succinate uptake with the extrusion of propionate and $CO_2$. *a* Succinate uptake system; *b* succinate propionyl-CoA: CoA transferase; *c* methylmalonyl-CoA mutase and methylmalonyl-CoA epimerase; *d* methylmalonyl-CoA decarboxylase; *e* ATPase

the only energy source for ATP synthesis. The bacteria have a $Na^+$-stimulated ATPase in their membrane that uses the $Na^+$ gradient established by the decarboxylase directly for ATP synthesis (Hilpert et al. 1984).

The properties of this $Na^+$ ATPase are strikingly homologous to $H^+$-translocating ATPases of the $F_1F_0$ type (Laubinger and Dimroth 1987, 1988, 1989). Like the ATPase of *Escherichia coli*, that of *P. modestum* consists of a losely membrane-bound $F_1$ moiety that harbors the catalytic sites for ATP synthesis/hydrolysis, and a more firmly membrane-bound $F_0$ part that is responsible for the cation conduction through the membrane. The subunit composition of the two enzymes is also homologous: each $F_1$ portion consists of five subunits ($\alpha$, $\beta$, $\gamma$, $\delta$, $\varepsilon$), and each $F_0$ sector contains three subunits (a,b,c). Each subunit c covalently binds dicyclohexyl carbodiimide, thereby inactivating the ATPase activity. Other inhibitors of the *E. coli* ATPase, e.g., venturicidin, tributyltin chloride, or azide were also inhibitory to the *P. modestum* enzyme. The close relationship between these ATPases was further supported by immunological and sequencing studies. Antibodies raised against the $\alpha$- and $\beta$-subunits of the *E. coli*

ATPase reacted with the corresponding subunits of the *P. modestum* enzyme, while no crossreactivities of other subunits were detectable (Laubinger et al. 1990). In the two $\beta$-subunits, 69% of the residues were identical (Amann et al. 1988). The $F_1F_0$ ATPase complex, but not $F_1$ alone, was specifically activated by $Na^+$ ions. This activation was lost upon dissociation of $F_1$ from $F_0$ and recovered upon its reconstitution with $F_0$, thus indicating that the $Na^+$ binding site is located on the $F_0$ moiety, not on $F_1$ (Laubinger and Dimroth 1987, 1988).

The function of the ATPase as a primary $Na^+$ pump (the reversal of the physiological reaction) was firmly secured by reconstitution experiments (Laubinger and Dimroth 1988). As shown in Fig. 6, sodium ions were pumped into proteoliposomes containing the ATPase in response to ATP hydrolysis. The transport was electrogenic, and the transport rate increased about fourfold after dissipation of the rate-determining membrane potential with either valinomycin (in presence of $K^+$) or the uncoupler carbonylcyanide *m*-chloro phenylhydrazone (CCCP). Stimulation rather than inhibition of $Na^+$ transport by the uncoupler firmly substantiates the functioning of the ATPase as a primary $Na^+$ pump and excludes the possibility of an $Na^+$ transport by the combined action of an

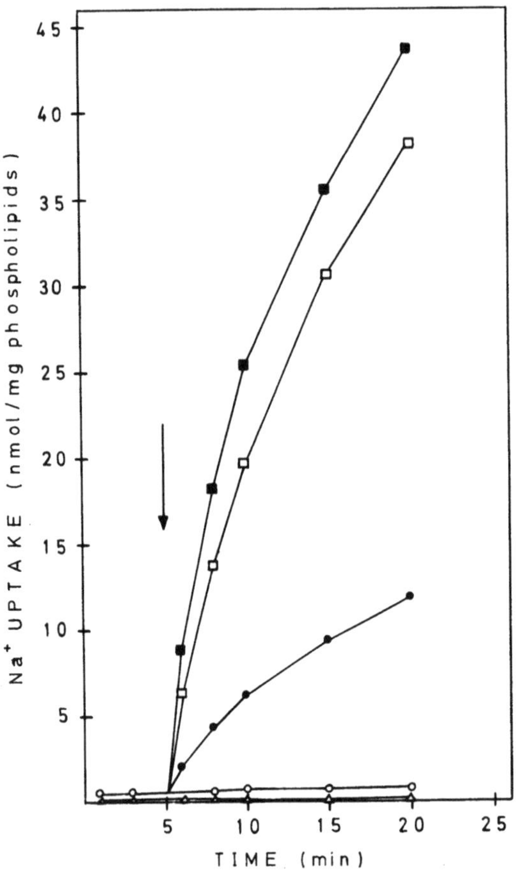


**Fig. 6.** Effect of ionophores on the kinetics of $Na^+$ uptake into reconstituted proteoliposomes containing the $F_1F_0$ ATPase of *P. modestum*. The addition at ATP is marked by the *arrow*; ionophores were contained as follows: (●) control; (■) 30 μM valinomycin; (□) 50 μM carbonylcyanide *m*-chlorophenylhydrazone; (△) 50 μM monensin; (○) control in the absence of ATP (Laubinger and Dimroth 1988)

$H^+$-translocating ATPase and an $Na^+/H^+$ antiporter. Amiloride, an inhibitor of such an antiporter, was without effect on the $Na^+$ transport, thus further confirming the conclusion that the ATPase acts as a primary $Na^+$ pump (Laubinger and Dimroth 1989).

The most pronounced difference between the ATPases of *E. coli* and *P. modestum* is the use of $H^+$ or $Na^+$, respectively, as the coupling ion. However, even in this respect, partial homology exists, because the *P. modestum* ATPase switches from $Na^+$ to $H^+$ translocation at low (< 1 mM) $Na^+$ concentrations (Laubinger and Dimroth 1989). Similar $Na^+$ concentrations (0.2–0.8 mM) yielded half maximal rates of $Na^+$ transport and ATP hydrolysis and reduced the rate of proton transport by one-half. We conclude therefore that the ATPase switches between $Na^+$ and $H^+$ translocation in response to the $Na^+$ concentration applied and further that the mechanism for the translocation of the two different cations is the same. Also congruent with this conclusion are recent results obtained with hybrid enzymes (Laubinger et al. 1990). Reconstituted proteoliposomal systems consisting of $F_0$ from *P. modestum* and $F_1$ from either *E. coli* or the thermophilic bacterium PS3 were capable of either $H^+$ or $Na^+$ pumping, and $Na^+$ prevented $H^+$ pumping at about the same concentration (> 1 mM) as in the reconstituted system with the homologous *P. modestum* ATPase. The ATPase of *E. coli* was on the other hand unable to catalyze $Na^+$ translocation. The $F_0$ part thus clearly defines the specificity for the coupling cation, and the ion translocation through $F_0$ apparently triggers ATP synthesis within the $F_1$ moiety, independent of whether the translocated cation is $Na^+$ or $H^+$. The change in cation specificity in $F_0$ could be caused by relatively minor alterations at the cation binding site, not significantly affecting the recognition site for $F_1$, as suggested from the formation of functional hybrids described above.

These findings have implications on the ATP synthesis/hydrolysis mechanism in general. Thus, any proposal in which $H^+$ ions play a specific role, none which could be performed by another cation as well, is unlikely to be correct. One such proposal by Mitchell (1974) assumes a direct participation of the vectorial protons in the chemistry of ATP synthesis at the catalytic site. The protons are supposed to flow down the electrochemical gradient through $F_0F_1$ into the catalytic site, where they attack one of the oxygen atoms of phosphate, forming $H_2O$ and $H_2PO_3^+$, which can directly react with ADP forming ATP. As this mechanism is obviously restricted to protons as coupling ions, it is disfavored by the discovery of an $Na^+$-coupled ATPase. We rather suppose in accord with the model of Boyer (1975) that binding of the cations to specific binding sites on $F_0$ from the side of high electrochemical potential triggers a conformational change which exposes the cations to the other surface of the membrane and brings about ATP synthesis on the $F_1$ moiety of the enzyme complex. After dissociation of the cations, the enzyme returns to its original conformation. It is also apparent that the cation conduction through $F_0$ cannot proceed via a network of a hydrogen-bonded chain ("proton wire"), but more likely requires specific binding sites. As shown for crown ethers, such sites could bind either $Na^+$ or $H_3O^+$ (Boyer 1988), indicating that $H_3O^+$ rather than $H^+$ may be the transported species.

### *2.3 Sodium as Coupling Ion in Other Bacteria*

Certain marine bacteria, such as *Vibrio alginolyticus*, possess a respiratory $Na^+$ pump which uses the free energy of NADH oxidation by ubiquinone to pump $Na^+$ ions out of the cell (Tokuda and Unemoto 1982). The $Na^+$ ion gradient generated by this pump may have several different physiological functions, such as the following: (1) the bacteria are nonmotile in the absence of $Na^+$ and acquire mobility in its presence, suggesting the operation of an $Na^+$-dependent flagellar motor in *V. alginolyticus* (Dibrov et al. 1986a); (2) several transport systems were shown to be $Na^+$-dependent (Tokuda and Unemoto 1982); and (3) the ATP level of bacterial cells increased after application of an $Na^+$ pulse, which could indicate the operation of an $Na^+$-motive ATPase (Dibrov et al. 1986b).

Most methanogenic bacteria form methane from $H_2$ and $CO_2$ and couple this exergonic reaction with the phosphorylation of ADP. While the overall process requires $Na^+$ (Perski et al. 1982), this alkali ion is apparently not directly involved in ATP synthesis, which appears to be coupled to an electrochemical gradient of protons (Müller et al. 1988a,b; Kaesler and Schönheit 1989). By comparing the effect of $Na^+$ on methane formation from different substrates and substrate combinations, it could be shown that $Na^+$ is specifically involved in the methylene/methyl-group interconversion (Müller et al. 1988a). The unfavorable oxidation of methanol to the formal redox level of formaldehyde is driven by $Na^+$ uptake into the cells. These $Na^+$ ions may recycle at the expense of $\Delta\tilde{\mu}H^+$ by the $Na^+/H^+$ antiporter. The reverse reaction has been shown to proceed by $Na^+$ extrusion from the cells through a primary $Na^+$ pump (Müller et al. 1988b). It has been reported that methanogens contain still another $Na^+$ pump that is driven by the conversion of formaldehyde to $CO_2$ and 2 $H_2$ (Kaesler and Schönheit 1989). Based on thermodynamics, the most likely site for this $Na^+$ pump is the formylmethanofuran dehydrogenase reaction [Eq. (3)]:

$$\begin{aligned}&\text{formylmethanofuran} + n\ Na^+_{in} + H_2O\\ &\rightleftharpoons CO_2 + H_2 + \text{methanofuran} + n\ Na^+_{out}.\end{aligned} \qquad (3)$$

In methanogenesis from $CO_2$ and $H_2$, the two $Na^+$-linked oxidoreductases could be coupled: $Na^+$ ions could be taken up to drive the endergonic reduction of $CO_2$ and exported again by the exergonic reduction of methylene to methyl groups. The formation of acetate by acetogenic bacteria involves a similar reduction of $CO_2$ to methyl groups, as in methanogens. Evidence has been presented that the reduction of methylenetetrahydrofolate to methyltetrahydrofolate is $Na^+$-dependent and may be coupled to $Na^+$ extrusion (Heise et al.1989). The dehydrogenases performing the methylene/methyl group interconversions could thus be analogous $Na^+$ pumps in methanogenic and acetogenic bacteria.

Sodium ions play a vital role in cell volume regulation by *Mycoplasma gallisepticum* (Shirvan et al. 1989a,b). These cells possess a $Na^+$-stimulated ATPase, suggested to function as a primary $Na^+$ pump. The ATPase was stimulated by $Na^+$ threefold at pH 8.5, but only very little at pH 5.5. These data are remeniscent to recent results obtained with the *P. modestum* ATPase (Laubinger and Dimroth

1989); they could indicate a switch between $Na^+$ and $H^+$ translocation in response to the concentrations of these two cations.

Alkalophilic bacteria are confronted with the problem of a low $\Delta\tilde{\mu}H^+$ due to a reversed proton gradient (acid inside), that may be insufficient to drive endergonic membrane reactions (Krulwich et al. 1988). An attractive hypothesis is therefore that these processes are coupled to the larger $\Delta\tilde{\mu}Na^+$. This hypothesis was shown to be true for the uptake of many solutes (Krulwich et al. 1982) and for the driving of the flagellar motor (Hirota et al. 1981). It was further demonstrated that pH homeostasis by these cells is $Na^+$-dependent and accomplished by the action of an $Na^+/H^+$ antiporter (Krulwich et al. 1988). In view of these findings and the existence of an $Na^+$-translocating ATPase in *P. modestum*, it was likely that ATP synthesis in alkalophilic bacteria was also driven by $\Delta\tilde{\mu}Na^+$. To investigate the ATP synthesis mechanism, we have isolated the ATPase of *Bacillus alcalophilus* (unpubl. results). The enzyme is of the $F_1F_0$ type, showing the characteristic pattern of eight different subunits on SDS-gel electrophoresis. The purified ATPase was not stimulated by $Na^+$ ions, and performed proton, not sodium ion pumping after reconstitution in proteoliposomes. In addition, the proteoliposomes catalyzed a $ATP/^{32}P_0$ exchange, that was uncoupler-sensitive but not inhibited by monensin. These results clearly indicate that ATP synthesis in *B. alcalophilus* is coupled to proton movement, and not to $Na^+$ movement as in *P. modestum*. Our results are completely in accord with recent studies on the mechanism of ATP synthesis by *Bacillus firmus* (Guffanti and Krulwich 1988). The difficulties inherent in the low $\Delta\tilde{\mu}H^+$ as driving force for ATP synthesis could be overcome, if the protons translocated by the respiratory chain would be guided to the ATPase without equilibrating with the protons in the bulk phase, as has been repeatedly proposed during the past two decades (for a review see Harold 1986). By this means, the actual driving force for ATP synthesis would increase considerably and would be essentially independent from the external pH, thus solving the bioenergetic problem of the alkalophilic bacteria.

It has been reported that *Bacillus FTU*, depleted of endogenous substrates, extruded some $Na^+$ ions by adding ascorbate and N,N,N′,N′-tetramethyl *p*-phenylenediamine (Semeykina et al. 1989). Additional $Na^+$ was extruded in the presence of the uncoupler CCCP. A much more dramatic effect of the uncoupler, however, was on intracellular $K^+$, from which the cells became essentially depleted. Subcellular vesicles of this *Bacillus* took up $Na^+$ on ascorbate oxidation, which was stimulated by CCCP or valinomycin. These data were interpreted as indication for a $Na^+$-motive terminal oxidase activity, but additional evidence is certainly required to prove this hypothesis. The same is true for a recent report, claiming the existence of an $Na^+$-motive NADH oxidase in *E. coli* (Avetisyan et al. 1989). In subcellular vesicles of these bacteria, grown at pH 8.6, but not in those grown at pH 7.6, $Na^+$ uptake was observed on NADH oxidation and in presence of CCCP. In these experiments, however, the internal $Na^+$ concentration remained well below the external $Na^+$ concentration applied.

## References


Amann, R., Ludwig, W., Laubinger, W., Dimroth, P. & Schleifer, K.H. (1988) Cloning and sequencing of the gene encoding the beta subunit of the sodium ion translocating ATP synthase of *Propionigenium modestum*. FEMS Microbiol. Lett. 56:253–260

Anantharam, V., Allison, M.J. & Maloney, P.C. (1989) Oxalate: formate exchange. J. Biol. Chem. 264:7244–7250

Avetisyan, A.V., Dibrov, P.A., Skulachev, V.P. & Sokolov, M.V. (1989) The $Na^+$-motive respiration in *Escherichia coli*. FEBS Lett. 254:17–21

Bergsma, J. & Konings, W.N. (1983) Citrate transport in membrane vesicles from *Bacillus subtilis*. Eur. J. Biochem. 134:151–156

Boyer, P.D. (1975) A model for conformational coupling of membrane potential and proton translocation to ATP synthesis and to active transport. FEBS Lett. 58:1–6

Boyer, P.D.(1988) Bioenergetic coupling to protonmotive force. Trends Biochem. Sci. 13:5–7

Buckel, W. (1986) Purification and reconstitution of glutaconyl-CoA decarboxylase from *Acidaminococcus fermentans*. Meth. Enzymol. 125:547–558

Buckel, W. & Liedtke, H. (1986) Specific cleavage of the sodium pump glutaconyl-CoA decarboxylase by primary alcohols. Eur. J. Biochem. 156:251–257

Buckel, W. & Semmler, R. (1983) Purification, characterization and reconstitution of glutaconyl-CoA decarboxylase. Eur. J. Biochem. 136:427–434

Däkena, P., Rohde, M., Dimroth, P. & Mayer, F. (1988) Oxaloacetate decarboxylase from *Klebsiella pneumoniae*: size and shape of the enzyme, and localization of its prosthetic biotin group by electron microscopic affinity labeling. FEMS Microbiol. Lett. 55:35–40

Dibrov, P.A., Kostyrko, V.A., Lazarova, R.L., Skulachev, V.P. & Smirnova, I.A. (1986a) $Na^+$-dependent motility and modes of membrane energization in *Vibrio alginolyticus*. Biochim. Biophys. Acta 850:449–457

Dibrov, P.A., Lazarova, R.L., Skulachev, V.P. & Verkhovskaya, M.L. (1986b) $Na^+$-coupled oxidative phosporylation in *Vibrio alginolyticus* cells. Biochim. Biophys. Acta 850:458–465

Dimroth, P. (1981) $Na^+$ transport by reconstituted oxaloacetate decarboxylase. J. Biol. Chem. 256:11974–11976

Dimroth, P. (1982a) The generation of an electrochemical gradient of sodium ions upon decarboxylation of oxaloacetate by the membrane-bound and $Na^+$-activated oxaloacetate decarboxylase from *Klebsiella aerogenes*. Eur. J. Biochem. 121:443–449

Dimroth, P. (1982b) The role of biotin and sodium in the decarboxylation of oxaloacetate by the membrane-bound oxaloacetate decarboxylase from *Klebsiella aerogenes*. Eur. J. Biochem. 121:435–441

Dimroth, P. (1986) Oxaloacetate decarboxylase from *Klebsiella aerogenes*. Meth. Enzymol. 125:530–540

Dimroth, P. (1987) Sodium ion transport decarboxylases and other aspects of sodium ion cycling in bacteria. Microbiol. Rev. 51:320–340

Dimroth, P. (1989) Mechanisms of sodium transport in bacteria. Phil. Trans. R. Soc. Lond. B 326:465–477

Dimroth, P. & Hilpert, W. (1984) Reversal of decarboxylation-dependent $Na^+$ pumps. Biochemistry 23:5360–5366

Dimroth, P. & Thomer, A. (1983) Subunit composition of oxaloacetate decarboxylase and characterization of the $\alpha$-chain as carboxyltransferase. Eur. J. Biochem. 137:107–112

Dimroth, P. & Thomer, A. (1986) Citrate transport in *Klebsiella pneumoniae*. Biol. Chem. Hoppe-Seyler 367:813–823

Dimroth, P. & Thomer, A. (1988) Dissociation of oxaloacetate decarboxylase and reconstitution of the active complex from the isolated subunits. Eur. J. Biochem. 175:175–180

Dimroth, P. & Thomer, A. (1990) Reconstitution of the $Na^+$-dependent citrate carrier of *Klebsiella pneumoniae*. J. Biol. Chem. 265:7221–7224

Guffanti, A.A. & Krulwich, T.A. (1988) $\Delta$pH-driven ATP synthesis in an alkalophilic *Bacillus*. J. Biol. Chem. 263:14748–14752

Harold, F.M. (1986) In: A study of bioenergetics. Freeman, New York, pp. 225–227

Heise, R., Müller, V. & Gottschalk, G. (1989) $Na^+$-dependence of acetate formation by *Acetobacterium woodii*. J. Bacteriol. 171:5473–5478

Hilpert, W. & Dimroth, P. (1983) Purification and characterization of a new sodium transport decarboxylase. Methylmalonyl-CoA decarboxylase from *Veillonella alcalescens*. Eur. J. Biochem. 132:579–587

Hilpert, W. & Dimroth, P. (1984) $Na^+$ transport by reconstituted methylmalonyl-CoA decarboxylase. Eur. J. Biochem. 138:579–583

Hilpert, W. & Dimroth, P. (1986) Sodium pump methylmalonyl-CoA decarboxylase from *Veillonella alcalescens*. Meth. Enzymol. 125:540–546

Hilpert, W., Schink, B. & Dimroth, P. (1984) Life by a new decarboxylation-dependent energy conservation mechanism with $Na^+$ as coupling ion. EMBO J. 3:1665–1670

Hirota, N., Kitada, M. & Imae, Y. (1981) Flagellar motors of alkalophilic *Bacillus* are powered by an electrochemical potential gradient of $Na^+$. FEBS Lett. 132:278–280

Hoffmann, A., Hilpert, W. & Dimroth, P. (1989) Carboxyltransferase activity of methylmalonyl-CoA decarboxylase. Eur. J. Biochem. 179:645–650

Kaesler, B. & Schönheit, P. (1989) The role of sodium ions in methanogenesis. Eur. J. Biochem. 184:223–232

Krulwich, T.A., Guffanti, A.A., Bornstein, R.F. & Hoffstein, J. (1982) Sodium requirement for growth, solute transport, and pH homeostasis in *Bacillus firmus* RAB. J. Biol. Chem. 257:1885–1889

Krulwich, T.A., Hicks, D.B., Seto-Young, D. & Guffanti, A.A. (1988) The bioenergetics of alkalophilic bacilli. CRC Crit. Rev. Microbiol. 16:15–36

Laubinger, W. & Dimroth, P. (1987) Characterization of the $Na^+$-stimulated ATPase of *Propionigenium modestum* as an enzyme of the $F_1F_0$ type. Eur. J. Biochem. 168:475–480

Laubinger, W. & Dimroth, P. (1988) Sodium ion translocating ATP synthase. Biochemistry 27:7531–7537

Laubinger, W. & Dimroth, P. (1989) Proton pumping by the $Na^+$-ATPase. Biochemistry 28:7194–7198

Laubinger, W., Deckers-Hebestreit, G., Altendorf, K. & Dimroth, P. (1990) $Na^+$ pumping by an $EF_1PF_0$ ATPase hybrid. Biochemistry 29:5458–5463

Laußermair, E., Schwarz, E., Oesterhelt, D., Reinke, H., Beyreuther, K. & Dimroth, P. (1989) Sequence of the $\beta$- and $\gamma$-subunits of oxaloacetate decarboxylase. J. Biol. Chem. 264:14710–14715

Mitchell, P. (1974) A chemiosmotic molecular mechanism for proton-translocating adenosine triphosphatases. FEBS Lett. 43:189–194

Müller, V., Blaut, M. & Gottschalk, G. (1988a) Electrochemical $Na^+$ gradient as driving force for methanol oxidation by *Methanosarcina barkeri*. Eur. J. Biochem. 172:601–606

Müller, V., Winner, C. & Gottschalk, G. (1988b) Electron-transport-driven sodium extrusion during methanogenesis. Eur. J. Biochem. 178:519–525

Perski, H.J., Schönheit, P. & Thauer, R.K. (1982) Sodium dependence of methane formation in methanogenic bacteria. FEBS Lett. 143:323–326

Radford, S.E., Laue, E.D., Perham, R.N., Miles, J.S. & Guest, J.R. (1987) Domains in pyruvate dehydrogenase complex. Biochem. J. 247:641–649

Schwarz, E., Oesterhelt, D., Reinke, H., Beyreuther, K. & Dimroth, P. (1988) Sequence of the $\alpha$-subunit of oxaloacetate decarboxylase. J. Biol. Chem. 263:9640–9645

Semeykina, A.L., Skulachev, V.P., Verkhovskaya, M.L., Bulygina, E.S. & Chemakov, K.M. (1989) $Na^+$-motive terminal oxidase in an alkalo- and halo-tolerant *Bacillus*. Eur. J. Biochem. 183:671–678

Shirvan, M.H., Schuldiner, S. & Rottem, S. (1989a) $Na^+$-cycle in *Mycoplasma gallisepticum* cell volume regulation. J. Bacteriol. 171:4410–4416

Shirvan, M.H., Schuldiner, S. & Rottem, S. (1989b) $Na^+$ extrusion by a primary $Na^+$ pump in *Mycoplasma gallisepticum*. J. Bacteriol. 171:4417–4424

Stephens, P.E., Darlison, M.G., Lewis, H.M. & Guest, J.R. (1983) The pyruvate dehydrogenase complex of *Escherichia coli*. Eur. J. Biochem. 133:481–489

Tokuda, H. & Unemoto, T. (1982) Respiration-dependent $Na^+$ pump in *Vibrio alginolyticus*. J. Biol. Chem. 257:10007–10014

Wifling, K. & Dimroth, P. (1989) Oxaloacetate decarboxylase of *Salmonella typhimurium*. Arch. Microbiol. 152:584–588

# Phosphorylative Electron Transport Without Quinone

A. KRÖGER, I. SCHRÖDER, B. KREMS, and O. KLIMMEK[1]

## 1 Introduction

The respiratory quinones are considered as essential components of bacterial and mitochondrial phosphorylative electron transport (Kröger 1977; Kröger et al. 1986). They serve as diffusible redox mediators between membrane-integrated dehydrogenases and electron transfer complexes such as the cytochrome $b/c_1$-complex, or certain reductases which react with $O_2$ or other terminal acceptors. In addition, quinones are postulated to be involved in the generation of the electrochemicalproton potential ($\Delta\tilde{\mu}_H$) across the respiratory membrane (von Jagow et al. 1986). The known respiratory systems that lack quinones are aerobic. In these systems, the proton pumping cytochrome oxidase (Chan and Li 1990) is thought to be responsible for generating the $\Delta\tilde{\mu}_H$. The systems of anaerobic respiration involve a quinone (Kröger 1977; Wissenbach et al. 1990) with the exception of sulfur respiration (Paulsen et al. 1986; Schröder et al. 1988). Therefore, it is of interest to find out how the $\Delta\tilde{\mu}_H$ is generated in this case. Here, we report on the most elaborate system of electron-transport-coupled phosphorylation with sulfur as acceptor, that is, the sulfur respiration of *Wolinella succinogenes*.

## 2 Results and Discussion

### *2.1 Energetics*

*W. succinogenes* can grow at the expense of either reaction (1) or (2).

$$HCO_2^- + \text{Fumarate} + H^+ \rightarrow CO_2 + \text{Succinate} \quad (1)$$
$$\Delta G_0' = -90 \text{ kJ/mol}$$
$$HCO_2^- + S^0 + H^+ \rightarrow CO_2 + H_2S \quad (2)$$
$$\Delta G_0' = -37 \text{ kJ/mol}$$

Both reactions are coupled to phosphorylation. Reaction (1) is catalyzed by the membrane of bacteria grown with fumarate and reaction (2) by that of bacteria grown with sulfur as the terminal acceptor (Macy et al. 1986). Generally, the free energy ($\Delta G_0'$) required for the synthesis of 1 mol ATP amounts to about 100 kJ in growing bacteria. Therefore, it is predicted that maximally 1 ATP/formate can be gained from reaction (1) and much less from reaction (2). Both reactions are

[1] Institut für Mikrobiologie der J.W. Goethe-Universität, Theodor-Stern-Kai 7, Haus 75A, W-6000 Frankfurt am Main, FRG

41. Colloquium Mosbach 1990
The Molecular Basis of Bacterial Metabolism

associated with the formation of a $\Delta\tilde{\mu}_H$, which is regarded as the coupling device between electron transport and phosphorylation according to the Mitchell theory. The measured phosphorylation yield with fumarate (2 ATP/3 formate) is in agreement with the shift of two protons per formate oxidized ($H^+/e = 1$) across the membrane and with a ratio of $H^+/ATP = 3$ (Table 1). The $\Delta\tilde{\mu}_H$ generated by reactions (1) and (2) are nearly the same, while the driving force ($\Delta E'$) for its generation is nearly equal to the $\Delta\tilde{\mu}_H$ with sulfur and more than twice the $\Delta\tilde{\mu}_H$ with fumarate as acceptor. Therefore, the $H^+/e$ ratio with sulfur should not exceed 0.5. With this value the ATP yield of reaction (2) is calculated to be 1 ATP/3 formate ($ATP/e = 1/6$).

**Table 1.** Properties of the anaerobic respiratory systems of *W. succinogenes*[a]

| *Reaction* | $\Delta E'$ (Volt) | $\Delta\tilde{\mu}_H$ | $H^+/e$ | ATP/e |
|---|---|---|---|---|
| a | 0.47 | 0.18 | 1 | 1/3 |
| b | 0.22 | 0.17 | 0.5 | 1/6 |

[a]The redox potential differences ($\Delta E'$) refer to the bacterial growth conditions with fumarate ($\Delta E_o'$) or sulfur (formate concentration 0.1 M, $p_{CO_2}$ and $p_{H_2S}$ = 0.1 bar). The values of $\Delta\tilde{\mu}_H$, $H^+/e$, and ATP/e were measured with fumarate as acceptor (Mell et al. 1986). The values of $H^+/e$ and ATP/e with sulfur were estimated from $\Delta E'$, $\Delta\tilde{\mu}_H$ (Wloczyk et al. 1989) and $H^+/ATP = 3$ (Brune et al. 1987).

### 2.2 Polysulfide as the Substrate

As the solubility of elemental sulfur in water (5 μg/l) is very small, soluble intermediates are expected to be formed which serve as direct acceptors of the bacterial electron transport. These intermediates are polysulfides ($S_n^=$), the various species of which undergo rapid dismutation in aqueous solution and therefore cannot be separated. Pentasulfide ($S_5^=$, pK = 5.7) is the most prominent species under the bacterial growth conditions (pH 8 and 4 mM sulfide) (Schwarzenbach and Fischer 1960). Up to 1 mM sulfur ($S^0$) can be solubilized at a velocity which is commensurate with that of the bacterial sulfide production [reaction (2); Krems 1989]. The solubilization reaction can be approximated by reaction (3).

$$S_8 + 2HS^- \rightarrow 2S_5^= + 2H^+. \quad (3)$$

When tetrathionate ($S_4O_6^=$) is added to a sulfide solution (pH 8) under anaerobic conditions, the solution becomes yellow and shows the same absorbance spectrum as a solution of elemental sulfur in sulfide (Klimmek 1990). This indicates that polysulfide is formed according to reaction (4):

$$4S_4O_6^= + 5HS^- \rightarrow 8S_2O_3^= + S_5^= + 5H^+ \quad (4)$$
$$\Delta G_0' = -219\ \text{kJ}.$$

The absorbance at 360 nm increases proportional to the amount of $S_4O_6^=$ added, indicating that a proportional amount of polysulfide is formed, in agreement with reaction (4). After the addition of *W. succinogenes* to the yellow solution, the absorbance decreases proportional with the amount of formate added. The absorbance decrease per mole of formate is equal to the absorbance increase per mole of $S_4O_6^=$ in the absence of bacteria. This indicates that the polysulfide formed by reaction (4) serves as the acceptor in the formate oxidation catalyzed by the bacteria [reaction (5)]:

$$4HCO_2^- + S_5^= + H^+ \rightarrow 4CO_2 + 5HS^- . \quad (5)$$

Actually the bacteria have been found to grow at the expense of the polysulfide formed according to reaction (4).

## 2.3 The Electron Transport Chain

The membrane fraction of *W. succinogenes* grown according to reaction (2) exhibits dehydrogenase activities with formate and sulfide. These enzymes have been isolated after solubilization of the membrane with Triton X-100 (Schröder et al. 1988). The formate dehydrogenase was indistinguishable from that isolated earlier from *W. succinogenes* grown with fumarate (Kröger et al. 1979). The sulfide dehydrogenase ($M_r$ 200 000) consisted essentially of an $M_r$ 85 000 polypeptide and contained equal amounts of free iron sulfide (120 μmol/g protein). Incorporation of the isolated enzymes into liposomes gave a preparation catalyzing reaction (2) at turnover numbers of the enzymes which were commensurate to those in the bacterial membrane (Table 2). In contrast, liposomes containing only one enzyme species were without electron transport activity. Thus the sulfide dehydrogenase was shown to act as a reductase using polysulfide as acceptor. A quinone was not

**Table 2.** Electron transport [reaction(2)] and enzyme activities of liposomes containing sulfur reductase and/or formate dehydrogenase[a]

| Liposomes containing | Formate → sulfur | Formate → DMN ($10^3$ $min^{-1}$) | Sulfide → DMN |
|---|---|---|---|
| Sulfur reductase | 0 | 0 | 1.4 |
| Formate DH | 0 | 13 | 0 |
| Formate DH + sulfur reductase | 1.1 | 10 | 4.3 |
| Formate DH + sulfur reductase + vit. $K_1$ | 1.2 | 9.6 | 5.7 |
| Bacterial membrane | 2.6 | 23 | 9.2 |

[a] The liposomes contained 0.1 g total enzyme protein per g phospholipid, and 3 g formate dehydrogenase per g sulfur reductase. The content of vitamin $K_1$ was 20 μmol/g phospholipid. The activities represent turnover numbers of the enzymes. The turnover numbers of electron transport (formate → sulfur) refer to sulfur reductase. DMN, 2,3-dimethyl-1,4-naphthoquinone.

required for restoring the electron transport activity [reaction (2)]. The addition of vitamin $K_1$, which was required for restoring the electron transport activity with fumarate [reaction (1)], did not cause stimulation of sulfur reduction by formate. The amount of MK which was associated with the enzymes inserted into the liposomal membrane was negligible and insufficient to restore reaction (1).

The experiment shown in Table 3 was designed to get a clue to the mechanism of transfer of the reducing equivalents from formate dehydrogenase to sulfur (polysulfide) reductase. The electron transport activities with sulfur [reaction (2)] and fumarate [reaction (1)] were measured with three different liposomal preparations. The liposomes contained formate dehydrogenase, sulfur reductase, and fumarate reductase at a fixed ratio, as well as vitamin $K_1$. The three preparations differed in the ratio of the amount of phospholipid over total protein. The electron transport activity with sulfur was found to decrease when that ratio increased, while the activity with fumarate remained nearly unchanged.

**Table 3.** Electron transport activities as a function of the liposomal phospholipid/protein ratio[a]

| Phospholipid/ protein ratio (g/g) | Formate → Sulfur | Formate → Fumarate |
|---|---|---|
| | ($10^3$ min$^{-1}$) | |
| 10 | 1.1 | 5.5 |
| 20 | 0.5 | 4.3 |
| 50 | 0.2 | 4.5 |

[a] The liposomes contained formate dehydrogenase, sulfur reductase and fumarate reductase at a ratio 3:1:1, and 20 μmol vitamin $K_1$/g phospholipid. The activities are given as turnover numbers referring to sulfur reductase or fumarate reductase.

This result is interpreted to indicate that the enzymes do not form electron transport complexes within the membrane, but are distributed randomly and independently of one another. The electron transfer with sulfur requires diffusion and collision of the two enzymes. From the phospholipid/protein ratio and the diffusion coefficient of the enzymes ($10^{-10}$cm$^2$/s) it is calculated that the velocity of the electron transport with sulfur should be limited by the migration of the enzymes and, therefore, decrease with the dilution of the enzymes. In contrast, the electron transfer from formate dehydrogenase to fumarate reductase should not be diffusion-limited. In this case, the transfer is mediated by MK, the diffusion coefficient ($10^{-6}$cm$^2$/s) of which is four orders of magnitude greater than that of the enzymes.

### *2.4 Mechanism of Energy Transduction*

The scheme given in Fig. 1 represents a hypothetical attempt to explain the mechanism of generation of the $\Delta\tilde{\mu}_H$ by reaction (2) and of the energy transduction

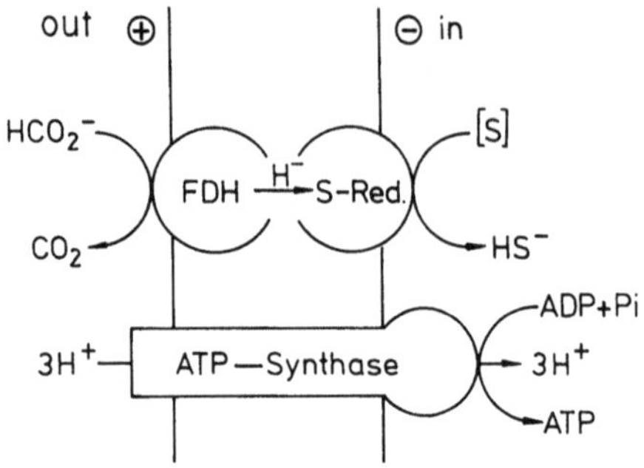


**Fig. 1.** Hypothetical mechanism of $\Delta\tilde{\mu}_H$ generation and the phosphorylation coupled to reaction (2). *FDH* Formate dehydrogenase; *S-Red.* sulfur reductase; [*S*], polysulfide

from the electron transport to phosphorylation. It is proposed that the $\Delta\tilde{\mu}_H$ results from the formal transport of a hydride ion across the membrane from formate to sulfur. Thus, per molecule of formate oxidized, a negative charge would disappear on the outside and appear on the inside of the membrane. Hydrolysis of the reaction products ($CO_2$ and $HS^-$) would lead to the external formation of one proton and to the internal consumption of one proton per molecule of formate oxidized. The mechanism which would explain the predicted $H^+/e$ ratio of 0.5 (Table 1) requires that the substrate site of formate dehydrogenase faces the bacterial outside, while that of the sulfur reductase should be oriented inside. The predicted orientation of the formate dehydrogenase was verified experimentally, while the sidedness of the sulfur reductase is not known so far. Also, the actual mechanism of transport of half a negative charge across the membrane per molecule of formate oxidized is not known. Its elucidation requires an understanding of the transfer of reducing equivalents between the two enzymes within the membrane. Also, the mechanism of reaction of sulfur reductase with its substrate (polysulfide) requires clarification.

## References

Brune, A., Spillecke, J. & Kröger, A. (1987) Correlation of the turnover number of the ATP synthase in liposomes with the proton flux and the proton potential across the membrane. Biochim. Biophys. Acta 893:499–507

Chan, S.I. & Li, P.M. (1990) Cytochrome c oxidase: understanding nature's design of a proton pump. Am. Chem. Soc. 29:1–12

Klimmek, O. (1990) Tetrathionat-Sulfid-Mischung als Substrat für das Wachstum von *Wolinella succinogenes*. Diploma Thesis, FB Biologie, J.W. Goethe-Universität Frankfurt am Main

Krems, B. (1989) Polysulfid als Substrat schwefel-reduzierender Bakterien. Diploma Thesis, FB Biologie, J.W. Goethe-Universität Frankfurt am Main

Kröger, A. (1977) Phosphorylative electron transport with fumarate and nitrate as terminal hydrogen acceptors. In: Microbial energetics. Cambridge Univ. Press, pp. 61–93

Kröger, A., Winkler, E., Innerhofer, A., Hackenberg, H. & Schägger, H. (1979) The formate dehydrogenase involved in electron transport from formate to fumarate in *Vibrio succinogenes*. Eur. J. Biochem. 94:465–475

Kröger, A., Paulsen, J. & Schröder, I. (1986) Phosphorylative electron transport chains lacking a cytochrome $bc_1$ complex. J. Bioenerg. Biomembr. 18:225–234

Macy, J.M., Schröder, I., Thauer, R.K. & Kröger, A. (1986) Growth of *Wolinella succinogenes* on $H_2S$ plus fumarate and on formate plus sulfur as energy sources. Arch. Microbiol. 144:147–150

Mell, H., Wellnitz, C. & Kröger, A. (1986) The electrochemical proton potential and the proton/electron ratio of the electron transport with fumarate in *Wolinella succinogenes*. Biochim. Biophys. Acta 852:212–221

Paulsen, J., Kröger, A. & Thauer, R.K. (1986) Energy-driven reverse of electron transport in the catabolism of *Desulfuromonas acetoxidans*. Arch. Microbiol. 144:78–83

Schröder, I., Kröger, A. & Macy, J.M. (1988) Isolation of the sulphur reductase and reconstitution of the sulphur respiration of *Wolinella succinogenes*. Arch. Microbiol. 149:572–579

Schwarzenbach, G. & Fischer, A. (1960) Die Acidität der Sulfane und die Zusammensetzung wässriger Polysulfidlösungen. Helv. Chim. Acta 43:1365–1390

von Jagow, G., Link, T.A. & Ohnishi, T. (1986) Organization and function of cytochrome b and ubiquinone in the cristae membrane of beef heart mitochondria. J. Bioenerg. Biomembr. 18:157–179

Wissenbach, U., Kröger, A. & Unden, G. (1990) The specific functions of MK and DMK in the anaerobic respiration with fumarate, demethylsulfoxide trimethylamine N-oxide and nitrate of *E. coli*. Arch. Microbiol. 154:60–66

Wloczyk, C., Kröger, A.; Göbel, T., Holdt, G. & Steudel, R. (1989) The electrochemical proton potential generated by the sulphur respiration of *Wolinella succinogenes*. Arch. Microbiol. 152:600–605

# Regulation of Gene Expression by Oxygen in *Escherichia coli*

J.R. Guest, J. Green, S. Spiro, C. Prodromou, and A.D. Sharrocks[1]

## 1 Introduction

*Escherichia coli* can grow under aerobic and anaerobic conditions deriving energy from different respiratory and fermentative processes. The former involve membrane-associated proton-translocating electron-transport pathways which generate utilizable energy by coupling oxidation of the growth substrate to the reduction of an exogenous electron acceptor such as oxygen, nitrate or fumarate. This contrasts with the fermentative processes where redox-balanced dismutations of the substrate generate energy by substrate-level phosphorylation. In the absence of oxygen, aerobic respiratory metabolism is replaced by anaerobic respiration or fermentation, depending on the availability of alternative electron acceptors. These changes are accompanied by altered patterns of protein synthesis reflecting the repression and derepression of the corresponding enzymes. It is also apparent that the regulatory mechanisms ensure that the most energetically favourable process is expressed in a specific environment. Thus, oxygen is the preferred electron acceptor relative to nitrate, and nitrate is used in preference to fumarate. In recent years transcriptional regulators have been identified for two networks of oxygen-regulated gene expression. One is the two-component sensor-regulator, ArcAB, which represses many aerobic enzymes under anaerobic conditions, the other is FNR, the transcriptional activator which is essential for the anaerobic induction of a variety of anaerobic functions. The properties of these regulatory systems will be discussed, particularly with respect to the effects of anaerobiosis on the citric acid cycle (CAC).

## 2 Results and Discussion

### *2.1 Central Pathways of Aerobic and Anaerobic Carbon Catabolism*

In *E. coli* the CAC is an inducible pathway which is most highly induced under conditions that make the greatest demands on its dual catabolic and anabolic functions, i.e. during aerobic growth on a non-fermentable substrate in media lacking amino acids and other preformed biosynthetic intermediates. It is subject to catabolite repression, anaerobic repression and end-product repression. Early studies showed that most of the CAC enzymes are repressed by glucose and further

[1]The Krebs Institute, Department of Molecular Biology and Biotechnology, University of Sheffield, Western Bank, Sheffield, S10 2TN, United Kingdom

41. Colloquium Mosbach 1990
The Molecular Basis of Bacterial Metabolism

repressed by anaerobiosis, the notable exceptions being succinate dehydrogenase and fumarase (but see below). Under aerobic conditions most substrates including growth-limiting amounts of glucose, are totally oxidized to $CO_2$ via the CAC (Fig. 1a), and an abundance of reducing equivalents are produced to energize the proton-translocating machinery. The net yield of ATP equivalents is approximately 22 per mol of glucose. The recovery from catabolite repression is mediated by cAMP which accumulates in the absence of a preferred substrate (e.g. glucose) and binds to the dimeric cAMP receptor protein (CRP). The resulting complex then activates the transcription of genes that are normally subject to catabolite repression, by interacting with specific sites in the promoter regions of the corresponding genes. The sequences of most of the CAC genes contain such CRP sites in their promoters (Miles and Guest 1987).

Under anaerobic conditions, or even aerobically in the presence of excess glucose, the CAC is transformed into a branched or non-cyclic pathway, due to the severe repression of the 2-oxoglutarate dehydrogenase complex (Fig. 1b). Carbon then flows at a much reduced rate through an oxidative route leading to 2-oxoglutarate (primarily for biosynthetic purposes), and through alternative reductive routes leading to the minor fermentation product, succinate. Under anaerobic conditions the pyruvate dehydrogenase complex is partially repressed and inactivated, whereas pyruvate formate-lyase is induced and activated, and as shown in Fig. 1b, glucose is converted into the major products of the mixed acid fermentation: ethanol; acetate; $H_2$; $CO_2$; and smaller amounts of succinate and formate (Clark 1989). Lactate is only produced under conditions of high pyruvate accumulation or high acidity, and the net yield of ATP is 3 per mol of glucose.

Also under anaerobic conditions a variety of compounds such as fumarate, nitrate, nitrite, trimethylamine *N*-oxide, dimethylsulphoxide and tetrahydrothiophene 1-oxide function as electron acceptors for specific terminal reductases. These are coupled through a quinone cofactor to specific primary dehydrogenases for substrates such as, glycerol 3-phosphate, lactate, formate, NADH and $H_2$. A variety of anaerobic respiratory systems can thus be induced for metabolizing both fermentable and non-fermentable substrates (Lin and Kuritzkes 1987; Stewart

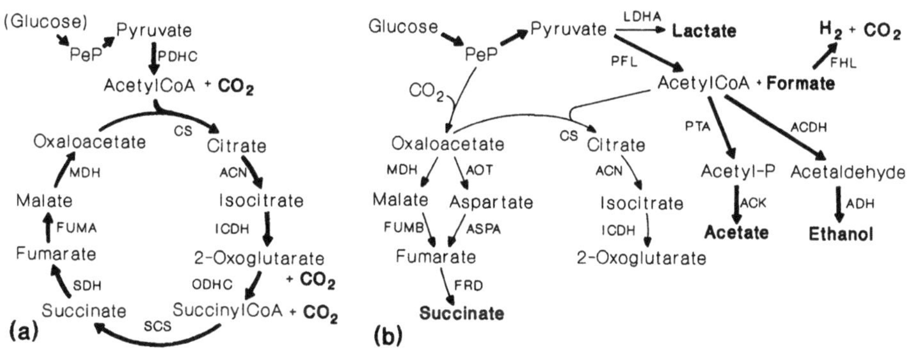


**Fig. 1a,b.** Central pathways of carbon metabolism under aerboic (**a**) and anaerobic (**b**) conditions. The major and minor routes are denoted by the relative thicknesses of the *arrows*

1988). Substrate oxidation is generally incomplete, presumably because the CAC is still repressed under these conditions.

### 2.2 The Discovery of Two Oxygen-Dependent Transcriptional Regulators

Our understanding of the two oxygen-dependent regulators, ArcA and Fnr, has in each case emerged from studies on the interconversion of succinate and fumarate. It has long been known that *E. coli* possesses two membrane-bound flavoprotein complexes which catalyze this interconversion but are specifically adapted in their kinetic properties and regulation, for succinate oxidation or fumarate reduction (Hirsch et al. 1963; Spencer and Guest 1974). Thus, succinate dehydrogenase (SDH) functions in the CAC donating reducing equivalents to the aerobic respiratory chaïn, whereas fumarate reductase (FRD) functions in the anaerobic reductive route as the terminal reductase of an anaerobic respiratory chain (Fig. 1b). The FRD system contributes to the efficiency of the mixed acid fermentation by accepting reducing equivalents and driving proton translocation, and it can also provide sufficient energy to allow *E. coli* to grow on non-fermentable substrates, such as glycerol, lactate, formate or $H_2$, provided that fumarate is supplied.

During the isolation of mutants lacking fumarate reductase a pleiotropic class was detected and designated *fnr* to denote a combined deficiency in *f*umarate and *n*itrate *r*eduction (Lambden and Guest 1976). The *fnr* gene was located at 29 min (Fig. 2), and it was subsequently shown to encode a transcriptional activator, FNR, closely resembling CRP (or CAP), the cAMP receptor protein or catabolite activator protein (Shaw et al. 1983). Thus, FNR activates the expression of a family of anaerobic genes in response to anoxia. It can also function as an anaerobic repressor in regulating both its own synthesis and the *ndh* gene. However, the mechanism by which it senses and responds to the presence or absence of oxygen is not understood.

The ArcAB system was identified by Lin and coworkers, who used strains containing an *sdh-lacZ* fusion to detect mutations that prevented the normal anaerobic repression of succinate dehydrogenase (Iuchi and Lin 1988; Iuchi et al. 1989a). The corresponding genes were designated *arcA* (0 min) and *arcB* (69 min) to denote defects in *a*erobic *r*espiration *c*ontrol (Fig. 2). The mutants were found

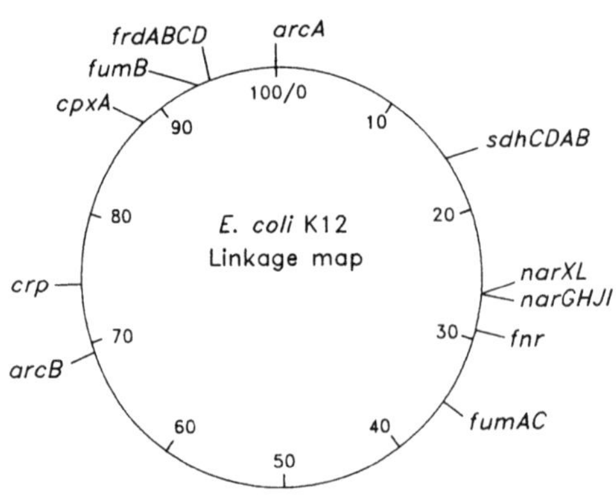


**Fig. 2.** Map locations of relevant transcriptional regulators and target genes

to lack the ability to repress a wide variety of enzymes of aerobic metabolism during anaerobic growth. These included enzymes of the CAC (except fumarase), the glyoxylate 'shunt', the fatty acid degradative pathway, the aerobic respiratory chain (cytochrome *o* oxidase) and several important flavoprotein dehydrogenases (e.g. the aerobic glycerol 3-phosphate dehydrogenase and the L-lactate and D-amino acid dehydrogenases), but the fermentative enzymes were unaffected.

The properties of the *arcA* mutants served to extend the pleiotropic phenotypes attributed to mutations in a gene previously designated *dye*, *sfrA*, *seg*, *fexA*, *msp* and *cpxC*. For example, *dye* mutations were known to increase sensitivity to toluidine blue, to prevent conjugation by loss of F-pili and to increase the synthesis of several outer and inner membrane proteins (Buxton et al. 1983). Indeed, two of the latter can now be recognized as the flavoprotein and iron-sulphur protein subunits of SDH. Furthermore, the *dye* gene (*arcA*) had already been cloned, sequenced and shown to encode a cytoplasmic protein ($M_r$ 27346) which is homologous to OmpR, the transcriptional regulatory component of the *ompR*-*envZ* system controlling porin synthesis (Drury and Buxton 1985). Further details of the ArcAB system are discussed in Section 2.3 below.

The synthesis of fumarase appeared to be relatively unresponsive to anaerobiosis. However, it is now known that *E. coli* contains three fumarase genes, *fumA*, *B* and *C* (Fig. 2). The *fumA* and *fumB* genes encode a homologous pair of novel dimeric iron-sulphur proteins, FUMA being an aerobic fumarate hydratase and FUMB an anaerobic malate dehydratase (Woods et al. 1988). The third gene, *fumC*, encodes a typical tetrameric fumarate hydratase (FUMC) that is expressed under aerobic and anaerobic conditions. Recent studies have now shown that FUMA is a CAC enzyme whose expression is regulated by CRP and ArcA, whereas FUMB is regulated by FNR but not ArcA (R. Roberts and J. Guest, unpublished). Thus, FUMA and SDH resemble each other in being CRP- and ArcA-regulated members of the CAC, whereas FUMB and FRD are their FNR-regulated anaerobic counterparts that function in the reductive branch of the anaerobic CAC (Fig. 1). Interestingly, aspartase and possibly asparaginase II, are under dual control by CRP and FNR. This arrangement presumably prevents oxidation of the amino acids in the presence of a preferred carbon source but facilitates their conversion to fumarate for use as an electron acceptor under anaerobic conditions. It is also apparent that cytochrome *o* oxidase is ArcA-regulated, whereas the cytochrome *d* oxidase, which functions at low $O_2$ tensions, is FNR-regulated (Iuchi and Lin 1988; Spiro and Guest 1990).

### *2.3 Aerobic Respiration Control*

The ArcAB system is a member of a growing family of two-component (sensor-regulator) signal-transducing proteins which control a diverse range of adaptive responses in bacteria (Gross et al. 1989; Stock et al. 1989). Each component has two important domains and the process of signal transduction is mediated by protein phosphorylation. Typically the sensor (or transmitter) is a transmembrane histidine protein kinase with an external sensor domain, which senses the environmental stimulus and induces autophosphorylation of the internal (*C*-terminal) transmitter domain. The signal is transmitted to the receiver domain of the cytoplasmic regulator (or receiver) by aspartate phosphorylation, and this in turn induces conformational changes in the *C*-terminal regulatory domain which affect DNA-binding activity or some other function. Several subfamilies have been identified and it is apparent that individual regulators are repressors, activators or both, and that some degree of cross-specificity for phosphorylation (cross-talk) can occur between different systems.

As yet little is known about ArcB, the sensor component, but it presumably senses anoxia or some consequential change in the redox or energy state of the membrane and converts the regulator, ArcA, into the form that represses the enzymes of aerobic metabolism (Fig. 3). More is known about CpxA, a typical sensor component which, with ArcA, is essential for expressing conjugal fertility functions (F-pilus synthesis), but not for the anaerobic repression of aerobic enzymes (Iuchi et al. 1989b). In this case it appears that CpxA senses the sexual state, and signal transduction converts ArcA into an activator for F-gene expression (Fig. 3). Thus, two quite distinct sensory kinases elicit different responses from a common regulator having two sets of target genes.

In view of the similarity between ArcA, OmpR and PhoB it seems likely that phosphorylated ArcA should likewise bind to short, non-palindromic sequences that are sometimes tandemly repeated in the regulatory regions of target genes. Unfortunately, a search for potential sites of ArcA-mediated anaerobic repression has not revealed any highly conserved sequence motifs. However, there is a poorly conserved motif in the non-coding regions of some of the CAC and related genes (Fig. 4). Comparable sequences occur in the *finP* and *traM* genes of the F-plasmid, but the significance of this motif is seriously challenged by the fact that similar sequences occur in genes which are not regulated by ArcA (e.g. *frdA*, *dld* and *ndh*). It is possible that ArcA regulation may affect transcription by a mechanism that does not involve site-specific binding to DNA, but it is hoped that a detailed molecular analysis of an ArcA-regulated promoter will reveal its mode of action, and a potential operator site may then emerge.

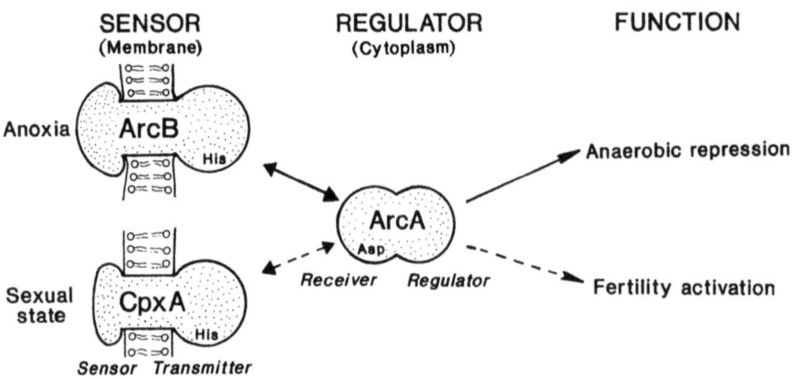


**Fig. 3.** Aerobic respiration control. Schematic representation of the components and domains of the ArcAB and CpxA-ArcA signal transducing systems that control anaerobic repression and fertility activation in *E. coli*

| | | |
|---|---|---|
| *aceEF* | -8 | TCAAtGGgACA |
| *lpd* | +61 | gCAAgGtAgCA |
| *gltA* $P_2$ | +34 | TCAAtGGAtCc |
| *sucAB* | -37 | gCAAgtGAACc |
| *sdhCDAB* $P_1$ | +1 | TCtcCGGAACA |
| *sdhDAB* $P_2$ | +68 | TggAaGaAACA |
| *fumA* | +2 | gCAACGGAACA |
| *mdhA* | ? | TaAACGGcAat |
| *aceBAK* | +3 | TgAACGaAAag |
| *cytb*$_{561}$ | (-10) | TtAACGGAAaA |
| Consensus | | TCAACGGAACA |
| *finP* | +3 | aCAtaGGAACc |
| *traM* | +174 | gCAAgGGAAaA |
| *frdA* | +7 | gCAgtGGAAtA |
| *dld* | (-192) | cCtACtGAACA |
| *ndh* | (-256) | cCtACcGAACg |

**Fig. 4.** Conserved sequences in ArcA-regulated genes. Sequences in aerobically expressed genes are shown above the consensus. Sequences found in other genes, including some that are not regulated by ArcA, are listed below the consensus. The position of the first base relative to the transcriptional start site is indicated

The preferential use of nitrate relative to fumarate or other anaerobic electron acceptors is also regulated by a related two-component system, NarX-NarL (Stock et al. 1989). In this case the sensor (NarX) detects exogenous nitrate (and molybdate), and initiates activation of the nitrate reductase operon (*narGHJI*) and repression of the fumarate reductase (*frdABCD*) and related operons via a response regulator (NarL), which probably binds to an appropriately positioned, partially palindromic target (Eiglmeier et al. 1989). This system in conjunction with FNR-mediated anaerobic activation establishes the hierarchy for the preferential use of terminal electron acceptors oxygen > nitrate > fumarate.

## 2.4 FNR-Regulated Gene Expression

FNR is a transcriptional activator (or repressor) of the functions listed in Table 1. It is synthesized under aerobic and anaerobic conditions but it becomes active as

**Table 1.** Functions regulated by FNR in *Escherichia coli*

| | |
|---|---|
| (a) *Anaerobic Activation* | Asparaginase II (*ansB*) |
| Fumarate reductase (*frdABCD*) | Aspartase (*aspA*) |
| Nitrate reductase (*narGHJI*) | Fumarase B (*fumB*) |
| Nitrite reductase (*nirB*) | Cytochrome *d* oxidase (*cyd*) |
| Cytochrome $c_{552}$ | Cytochrome $a_1$ |
| Dimethyl sulphoxide reductase (*dmsABC*) | Molybdate reductase |
| Trimethylamine N-oxide reductase (*tor*) | |
| Glycerol 3-phosphate dehydrogenase (*glpABC*) | (b) *Anaerobic Repression* |
| Pyruvate formate-lyase (*pfl*) | FNR (*fnr*) |
| Nickel transport (*hydC*) | NADH dehydrogenase II (*ndh*) |

a transcriptional regulator primarily under anaerobic conditions. The mode of action of FNR has been inferred from its similarity to CRP but there is considerable supporting evidence from biochemical and molecular-genetic studies. Thus, it is apparent that FNR is a sequence-specific DNA-binding protein which recognizes a partially palindromic target sequence via a helix-turn-helix structural motif. It is likewise assumed that the functional form is dimeric and that the active conformation is induced via a metal-ion prosthetic group or an unknown cofactor that signals the onset of anaerobiosis.

The CRP monomer has two domains, a nucleotide-binding domain and a *C*-terminal DNA-binding domain with the helix-turn-helix motif, $\alpha_E$-$\alpha_F$ (Fig. 5). Sequence alignment and secondary structural prediction indicates that FNR has essentially the same structure, except for a unique cysteine-rich extension at the *N*-terminus and some extra residues at the *C*-terminus (Shaw et al. 1983). The CRP-FNR family has recently been joined by three further proteins: FixK, a component of the regulatory cascade controlling $N_2$-fixation in *Rhizobium meliloti* (Batut et al. 1989); HlyX, a potential regulator of haemolysin synthesis in *Actinobacillus pleuropneumoniae* (J. MacInnes, unpublished); and FnrN, a regulator of microaerobic nitrogen fixation in *Rhizobium leguminosarum* (U. Priefer, unpublished). There is a high degree of sequence conservation, particularly in the DNA recognition helix ($\alpha_F$), and for the residues separating several elements of secondary structure (Fig. 5). The importance of the glycine residue between $\beta_5$ and $\beta_6$ in FNR has recently been confirmed by finding Gly96 → Asp substitutions in four independent *fnr* mutants (A. Sharrocks, unpublished). The residues in $\beta_7$ and $\alpha_C$ of CRP that interact with cAMP are not retained in the other proteins. However, the

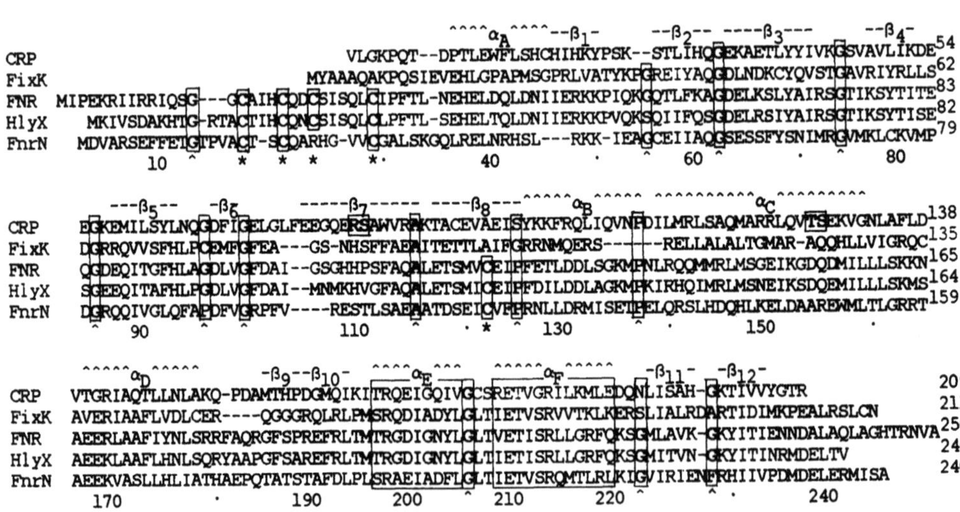


**Fig. 5.** Sequence alignments for the CRP-FNR family of transcriptional regulators. Secondary structures are shown for the nucleotide-binding ($\alpha_A$-$\alpha_C$) and DNA-binding ($\alpha_D$-$\beta_{12}$) domains of CRP (Weber and Steitz 1987). Residues with an important structural or functional significance are *boxed* and the *numbering* refers to FNR

cysteine cluster and the conserved Cys122 in FNR, HlyX and FnrN (Fig. 5) may be common features of regulators that respond to anoxia.

The FNR protein has been isolated from amplified sources by a variety of methods using immunological or DNA-binding assays to monitor the purification (Unden and Guest 1985; Trageser et al. 1990; J. Green, unpublished). The product is a monomeric protein having the properties listed in Table 2. It is susceptible to proteolytic removal of nine *N*-terminal residues during extraction from some hosts. However, recent studies with an in vitro deleted *fnr* gene have indicated that removing residues 2 to 9 has little effect on FNR activity. Attempts to isolate or generate a dimeric form have failed, as have attempts to demonstrate *site-specific* binding to target DNA: the monomeric form exhibits a non-specific affinity for DNA of $2 \times 10^{-7}$ M.

The key problems associated with the mode of action of FNR are the response to anoxia and the molecular basis of site-specific transcriptional activation and repression.

**Table 2.** Biochemical properties of FNR

| | |
|---|---|
| Apparent $M_r$ (crude extract) | 50 000–60 000 and 29 000 (sedimentation) |
| Apparent $M_r$ (pure) | 29 000 (gel filtration) |
| | 27 000 and 30 000 (SDS-PAGE) |
| | 30 000 (SDS-PAGE, 2-mercaptoethanol-treated) |
| Predicted $M_r$ | 27 965 |
| Absorption spectrum | $\lambda_{max}$ 280 nm (shoulder 420 nm, $\varepsilon$ = 1.1 $M^{-1}$ $cm^{-1}$) |
| Acid-labile sulphide | < 0.02 mol/mol |
| Titratable sulphydryl | 0.5–3.2 mol/mol (native) |
| | 3.2–3.7 mol/mol (denatured) |
| Metal ions | $Fe^{2+}$ plus $Fe^{3+}$, 0.02–1.10 mol/mol; $Zn^{2+}$, 0.10 mol/mol |
| | $Cu^{2+}$, $Mo^{6+}$, $Ni^{2+}$ and $V^{3+}$, < 0.01 mol/mol |
| Phosphate | None |

### *2.5 The Sensing of Anoxia by FNR*

A unique feature of FNR compared to CRP is the cluster of cysteine residues (Cys16, 20, 23, 29) plus Cys122. It is therefore reasonable to predict that these residues are involved in the response to anoxia, e.g. in disulphide-dithiol exchange reactions, interaction with a coeffector or provision of a metal binding site. Evidence for their importance comes from the inactivation that accompanies deletion of residues 3–30, or single Cys20 → Ser, Cys23 → Gly, Cys29 → Ala and Cys122 → Ala (but not Cys16 → Ala) substitutions (Spiro and Guest 1988; A. Sharrocks, unpublished). It may also be significant that a proportion of purified FNR has an apparent $M_r$ of 27 000 which becomes 30 000 upon treatment with 2-mercaptoethanol (Table 2) because this would be consistent with the presence of an intramolecular disulphide bond (J. Green, unpublished).

Considerable attention is now focussed on the probability that FNR contains an iron atom that is liganded to sulphydryl groups and responds to the redox state of the cell. Preparations of FNR isolated under aerobic and anaerobic conditions contain or bind up to 1.1 atoms of Fe per mol, the iron and sulphydryl contents being inversely related (J. Green, J.R. Guest, M. Trageser and G. Unden, unpublished). Indirect evidence for a role for iron (probably $Fe^{2+}$) in the anoxia-sensing pathway came from studies with the chelating agent *o*-phenanthroline, which mimics oxygen in inhibiting the anaerobic activation or repression of the respective *frd* or *ndh* promoters in growing bacteria (Spiro et al. 1989). In parallel studies Trageser and Unden (1989) showed that three (net) cysteine residues can be carboxymethylated in permeabilized cells, and that the rate of alkylation is higher in aerobic bacteria and in the presence of a chelating agent than in anaerobic bacteria. This suggests that a change in the binding or oxidation state of iron occurs under anaerobic conditions and these affect the accessibility of the cysteine sulphydryl groups. It is now important to understand the signal transducing system that induces this change, and to relate it directly via conformational change to the proposed DNA-binding and transcriptional regulatory functions of FNR.

### *2.6 The DNA-Binding Specificity of FNR*

The sequences recognized by dimeric DNA-binding proteins with helix-turn-helix motifs generally extend over about 22 bp and they are partial palindromes. A consensus sequence for the FNR site was originally detected by comparing the promoter regions of FNR-regulated genes (Spiro and Guest 1987a), and refined as more genes were sequenced (e.g. Eiglmeier et al. 1989). Confirmation was obtained from several types of experiment including: transfer of the FNR site of *nirB* to the *gal* promoter so as to render *gal* expression FNR-dependent, and studies with mutant FNR sites (Jayaraman et al. 1988, 1989); the identification of an FNR-regulated gene (*ndh*) by the presence of an FNR site (Spiro et al. 1989); and imposing FNR dependence upon the *melR* promoter by replacing its CRP site with a synthetic FNR site (Bell et al. 1989). The FNR site consensus resembles that of the CRP site,

FNR: 5′ A – A – **T T G A T** – – A – **A T C A A T** – – – 3′;
CRP: 5′ A A – **T G T G A** – – – – – – **T C A C A** – T T 3′.

Both have core motifs which are strongly conserved in each half-site. Studies with mutant sequences have shown that **T(A)** and **G(C)** at positions 5(18) and 7(16) are known to be essential for FNR specificity, whereas **G(C)** at both positions is essential for conferring CRP specificity.

The mechanism of sequence recognition by FNR has been explored by in vitro mutagenesis studies based on the sequence similarities in the DNA-binding domains of FNR and CRP and on the model for the CRP:DNA interaction. The X-ray structures of several protein:DNA complexes have shown that the binding specificity is conferred largely by hydrogen bonding between amino acid side chains in the DNA recognition helix ($\alpha_F$), and the base pairs that are exposed in the major groove of the double helix. In CRP there is indirect evidence that Arg180 and Glu181 form hydrogen bonds with the **GC** base pairs in the **TGTGA** motif as shown in Fig. 6. Furthermore, by aligning the DNA recognition helices of FNR and CRP it was correctly predicted that three substitutions in FNR (Val208 → Arg, Ser212 → Gly and Gly216 → Lys) should produce an altered FNR which is capable of

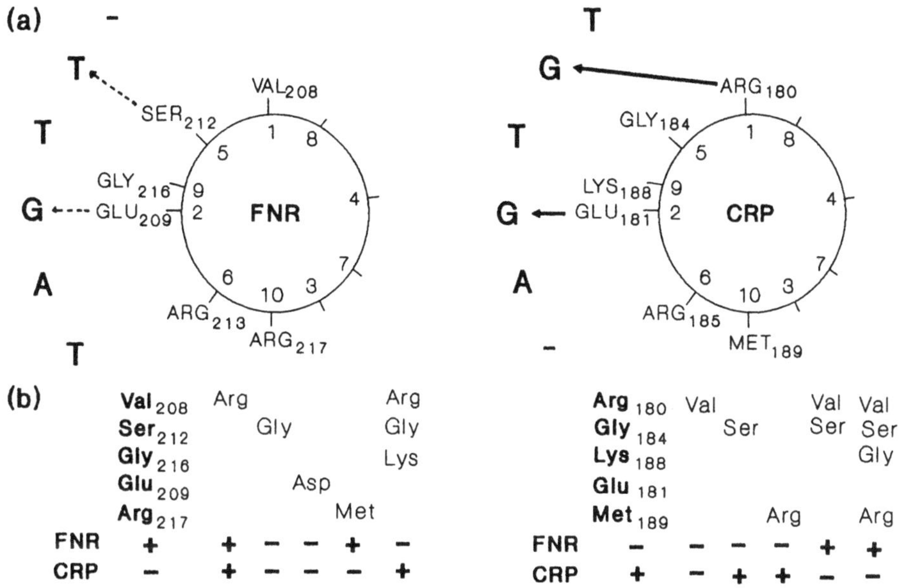


**Fig. 6a,b.** Site-directed mutagenesis of the DNA-binding domains of FNR and CRP. **a** End-on views of the DNA recognition helices ($\alpha_F$) with the cognate half-site consensus sequences and proposed interactions. **b** Amino acid substitutions and FNR or CRP activities of mutants (S. Spiro, R.E. Roberts, A.J. Bell, K.L. Gaston, S.J.W. Busby and J.R. Guest, unpublished)

activating CRP-dependent genes in response to anaerobiosis (Spiro and Guest 1987b). It is now known that a single substitution (Val208 → Arg) is sufficient to confer CRP specificity without loss of FNR specificity (Fig. 6). It is also known that Glu209 is essential, that Ser212 is responsible for discriminating between FNR and CRP sites, and that Arg217 is not essential (Fig. 6). It would thus appear that Glu209 and Ser212 are key residues in defining the binding specificity of FNR, presumably by interacting with the first **TA** and the **GC** base pairs in the **TTGAT** motif. The importance of Ser212 in contributing to the DNA-binding specificity is strengthened by its conservation in FNR, FixK, HlyX and FnrN, all of which appear to recognize the same target (Fig. 5).

Attempts to construct a CRP protein which activates FNR-dependent genes in response to cAMP have been less successful. Introducing a Gly184 → Ser substitution does not confer FNR specificity (Fig. 6). However, a weak but significant activity is obtained with an additional Arg180 → Val replacement, but further changes at Lys188 and Met189 to create the side-chain arrangement found in FNR afford no enhancement (Fig. 6). It would appear that further contacts at sites outside the recognition helices affect the overall functional efficacy of these regulators.

## 3 Conclusions

The regulators discussed here function primarily as the anaerobic activator of 'anaerobic' genes (FNR), or the anaerobic repressor of 'aerobic' genes (ArcA). However, FNR can be an anaerobic repressor, so it is conceivable that the ArcA could likewise be an anaerobic activator for some genes. It would also appear that these regulators do not control the expression of all oxygen-dependent genes in *E. coli*, notable exceptions being the glycolytic and fermentation genes. Indeed, it is known that environmental stresses such as anaerobiosis can alter the degree of DNA supercoiling, and that supercoiling affects the strengths of many promoters (Ni Briain et al. 1989). So it would appear that FNR and ArcAB represent specific regulatory systems operating within the framework of a general response to aerobic/anaerobic stress. Other specific regulators include OxyR, which activates a family of genes in response to the oxidative stress imposed by hydrogen peroxide (Christman et al. 1989).

Oxygen has a dramatic effect on gene expression in other bacteria, for example the photosynthetic apparatus of purple non-sulphur bacteria is induced at low oxygen tension, and the ability to fix nitrogen in free-living and symbiotic bacteria is only expressed under anaerobic or microaerobic conditions. Here, comparable systems and common themes are found; e.g. the two-component and CRP/FNR-like regulators, cysteine-rich clusters, metal-ion requirements and FNR sites (Spiro and Guest 1990). However, in no case is the oxygen-sensing or ultimate transcriptional switch understood, and these are the immediate goals of future research.

*Acknowledgements.* We are grateful to the SERC and the Leverhulme Trust for financial support, and we are indebted to Drs. S. Busby, J. MacInnes, U. Priefer, M. Trageser and G. Unden for allowing us to quote their unpublished work.

## References

Batut, K., Daveran-Mingot, M.L., David M., Jacobs, J., Garnerone, A.M. & Kahn, D. (1989) *fixK*, a gene homologous with *fnr* and *crp* from *Escherichia coli*, regulates nitrogen fixation genes both positively and negatively in *Rhizobium meliloti*. EMBO J. 8:1279–1286

Bell, A.I., Gaston, K.L., Cole, J.A. & Busby, S.J.W. (1989) Cloning of binding sequences for the transcription activators, FNR and CRP: location of bases involved in discrimination between FNR and CRP. Nucl. Acid. Res. 17:3865–3874

Buxton, R.S., Drury, L.S. & Curtis, C.A.M. (1983) Dye sensitivity correlated with envelope protein changes in dye (*sfrA*) mutants of *Escherichia coli* K12 defective in the expression of the sex factor F. J. Gen. Microbiol. 129:3363–3370

Christman, M.F., Stortz, G. & Ames, B.N. (1989) OxyR, a positive regulator of hydrogen peroxide-inducible genes in *Escherichia coli* and *Salmonella typhimurium*, is homologous to a family of bacterial regulatory proteins. Proc. Natl. Acad. Sci. USA 86:3484–3488

Clark, D.P. (1989) The fermentative pathways of *Escherichia coli*. FEMS Microbiol. Rev. 63:223–234

Drury, L.S. & Buxton, R.S. (1985) DNA sequence analysis of the *dye* gene of *Escherichia coli* reveals amino acid homology between Dye and OmpR proteins. J. Biol. Chem. 260:4236–4242

Eiglmeier, K., Honore, N., Iuchi, S., Lin, E.C.C. & Cole, S.T. (1989) Molecular genetic analysis of FNR-dependent promoters. Mol. Microbiol. 3:869–878

Gross, R., Arico, B. & Rappuoli, R. (1989) Families of bacterial signal-transducing proteins. Mol. Microbiol. 3:1661–1667

Hirsch, C.A., Rasminsky, M., Davis, B.D. & Lin, E.C.C. (1963) A fumarate reductase in *Escherichia coli* distinct from succinate dehydrogenase. J. Biol. Chem. 238:3770–3774

Iuchi, S. & Lin, E.C.C. (1988) *arcA*(dye), a global regulatory gene in *Escherichia coli* mediating repression of enzymes in aerobic pathways. Proc. Natl. Acad. Sci. USA 85:1888–1892

Iuchi, S., Cameron, D.C. & Lin, E.C.C. (1989a) A second global regulator gene (*arcB*) mediating repression of enzymes in aerobic pathways of *Escherichia coli*. J. Bacteriol. 171:868–873

Iuchi, S., Furlong, D. & Lin, E.C.C. (1989b) Differentiation of *arcA*, *arcB* and *cpxA* mutant phenotypes of *Escherichia coli* by sex pilus formation and enzyme regulation. J. Bacteriol. 171:2889–2893

Jayaraman, P.S., Gaston, K.L., Cole, J.A. & Busby, S.J.W. (1988) The *nirB* promoter of *Escherichia coli*: location of sequences essential for regulation by oxygen, the FNR protein and nitrite. Mol. Microbiol. 2:527–530

Jayaraman, P.S., Cole, J.S. & Busby, S.J.W. (1989) Mutational analysis of sequences at the FNR-dependent *nirB* promoter in *Escherichia coli*. Nucl. Acid. Res. 17:135–145

Lambden, P.R. & Guest, J.R. (1976) Mutants of *Escherichia coli* K12 unable to use fumarate as an anaerobic electron acceptor. J. Gen. Microbiol. 97:145–160

Lin, E.C.C. & Kuritzkes, D.R. (1987) Pathways of anaerobic electron transport. In: Neidhardt, F.C., Ingraham, J.L., Low, K.B., Magasanik, B., Schaechter, M. & Umbanger, H.E. (eds) *Escherichia coli* and *Salmonella typhimurium* cellular and molecular biology. Am. Soc. Microbiol., Washington, vol 1, pp. 201–221

Miles, J.S. & Guest, J.R. (1987) Molecular genetic aspects of the citric acid cycle of *Escherichia coli*. Biochem. Soc. Symp. 54:45–65

Ni Briain, N., Dorman, C.J. & Higgins, C.F. (1989) An overlap between osmotic and anaerobic stress responses: a potential role for DNA supercoiling in the coordinate regulation of gene expression. Mol. Microbiol. 3:933–942

Shaw, D.J., Rice, D.W. & Guest, J.R. (1983) Homology between CAP and Fnr, a regulator of anaerobic respiration in *Escherichia coli*. J. Mol. Biol. 166:241–247

Spencer, M.E. & Guest, J.R. (1974) Proteins of the inner membrane of *Escherichia coli*: changes in composition associated with anaerobic growth and fumarate reductase amber mutation. J. Bacteriol. 117:954–959

Spiro, S. & Guest, J.R. (1987a) Regulation and overexpression of the *fnr* gene of *Escherichia coli*. J. Gen. Microbiol. 133:3279–3288

Spiro, S. & Guest, J.R. (1987b) Activation of the *lac* operon of *Escherichia coli* by a mutant FNR protein. Mol. Microbiol. 1:53–58

Spiro, S. & Guest, J.R. (1988) Inactivation of the FNR protein of *Escherichia coli* by targeted mutagenesis in the *N*-terminal region. Mol. Microbiol. 2:701–707

Spiro, S. & Guest, J.R. (1990) FNR and its role in oxygen-regulated gene expression in *Escherichia coli*. FEMS Microbiol. Rev. 75:399–428

Spiro, S., Roberts, R.E. & Guest, J.R. (1989) FNR-dependent repression of the *ndh* gene of *Escherichia coli* and metal ion requirement for FNR-regulated gene expression. Mol. Microbiol. 3:601–608

Stewart, V. (1988) Nitrate respiration in relation to facultative metabolism in Enterobacteria. Mic. Rev. 52:190–232

Stock, J.B., Ninfa, A.J. & Stock, A.M. (1989) Protein phosphorylation of adaptive responses in bacteria. Microbiol. Rev. 53:450–490

Trageser, M. & Unden, G. (1989) Role of cysteine residues and metal ions in the regulatory functioning of FNR, the transcriptional regulator of anaerobic respiration in *Escherichia coli*. Mol. Microbiol. 3:593–599

Trageser, M., Spiro, S., Duchene, A., Kojro, E., Fahrenholz, F., Guest, J.R. & Unden, G. (1990) Isolation of intact FNR protein ($M_r$ 30000) of *Escherichia coli*. Mol. Microbiol. 4:21–27

Unden, G. & Guest, J.R. (1985) Isolation and characterization of the FNR protein, the transcriptional regulator of anaerobic electron transport in *Escherichia coli*. Eur. J. Biochem. 146:193–199

Weber, I.T. & Steitz, T.A. (1987) Structure of a complex of catabolite gene activator protein and cyclic AMP refined at 2.5Å resolution. J. Mol. Biol. 198:311–326

Woods, S.A., Schwartzbach, S.D. & Guest, J.R. (1988) Two biochemically distinct classes of fumarase in *Escherichia coli*. Biochim. Biophys. Acta 954:14–26

# Regulation of Gene Expression by Oxygen: Phototrophic Bacteria

S.S. TAREMI and B.L. MARRS[1]

## 1 Introduction

Pure cultures of certain phototrophic bacteria undergo striking changes in color in response to varying oxygen tensions. *Rhodobacter capsulatus* cells in anaerobic cultures are dark brown. A tiny amount of oxygen during or after growth results in deep reddish purple cultures, while cells grown in the presence of atmospheric oxygen are a light pastel shade of orange-pink. This variability has fascinated microbiologists for more than half a century (van Niel 1944). As our analytical abilities and understanding of genetics have increased, we have come to recognize these color changes as parts of an intricate metabolic response to a key environmental factor: oxygen.

Oxygen affects most living things at multiple levels, because it is both a useful and a potentially dangerous oxidant. In coping with this double-edged sword, many species have evolved mechanisms that respond to changes in oxygen concentration by turning genes on or off, but the first example of how this can be accomplished has only recently been described in molecular terms for one system (Storz et al. 1990), and other systems seems to differ. The phototrophic bacteria provide a different system in which we hope to be able to describe exactly how oxygen concentration-dependent signals are transduced into gene-regulating actions.

It has been demonstrated that the *puf* operon of *Rhodobacter capsulatus*, consisting of genes for pigment synthesis and pigment-binding proteins, is regulated in response to oxygen tension (Clark et al. 1984; Belasco et al. 1985). It now appears that this response is mediated by regulatory proteins, which preliminary findings suggest are yet another example of a so-called two-component regulatory system. Two-component regulatory systems, composed of a sensory and a regulatory protein, are found in a wide variety of bacterial systems that respond to environmental stimuli (Stock et al. 1989). If such a system is indeed involved in transducing the oxygen signal to the genome, the problem of describing how oxygen controls this specific example of gene expression will devolve to a study of what exactly this particular sensory protein senses.

In this chapter we shall review some of the evidence for control of gene expression by oxygen in the phototrophic bacteria, and we shall introduce some recent unpublished evidence from several laboratories that suggests that we may be closing in on an understanding of this phenomenon at the molecular level.

[1]Central Research and Development Department, E.I. du Pont de Nemours & Company, Inc., Experimental Station, P.O. Box 80173, Wilmington, DE 19880-0173, USA

41. Colloquium Mosbach 1990
The Molecular Basis of Bacterial Metabolism

## 2 Results and Discussion

### 2.1 *Review*

The first published model to address the mechanism of control of pigment synthesis by oxygen (Cohen-Bazire et al. 1957) suggested that since high light intensity and oxygen each repress synthesis, and each were also thought to oxidize the electron transport chain, perhaps one of the components of the electron transport system was the regulatory signal that controlled pigment synthesis. Subsequent attempts to find a redox component that became oxidized over the range of oxygen tensions which shut off pigment synthesis were unsuccessful (Schön and Drews 1968).

Marrs and Gest (1972) introduced a genetic approach to the question of whether or not a component of the electron transport system provided the regulatory signal controlling pigment synthesis. They studied the response of pigment synthesis to oxygen in two types of mutants of *Rhodobacter capsulatus*. In the first mutant both NADH and succinate dehydrogenases were inactivated, and pigment synthesis in this mutant strain became much more sensitive to oxygen than that of the wild-type parent. This result supported the notion that the flow of electrons into the electron transport system did play a role in the regulation of pigment synthesis. The second mutant, however, cast doubt on whether that role was to serve as a regulatory signal. The second mutant analyzed lacked the two major oxidases by which molecular oxygen oxidizes the electron transport system. The physiological effect of the mutations in the oxidase-negative mutant was profound, causing an almost total loss of the ability to consume oxygen or to grow by respiration, yet these cells showed a transitory inhibition of bacteriochlorophyll synthesis by oxygen very much like that in the wild type. Marrs and Gest summarized their observations by suggesting a model in which a regulatory factor was reduced by electrons from the electron transport chain but was not itself a component of the chain, since it was not dependent on the electron transport system oxidases for communication with oxygen.

When the genes for the photosynthetic apparatus were cloned (Marrs et al. 1981) and aligned with the genetic map (Taylor et al. 1983), it became possible to examine the transcription of those genes for oxygen-induced changes (Clark et al. 1984). It was immediately obvious that transcripts of some genes increased dramatically when oxygen tension was lowered from 20 to 2%, and that the genes affected were part of the *puf* operon. Much additional work by many laboratories has been required to understand the nature of the mRNA transcripts involved (reviewed in Scolnik and Marrs 1987), and many puzzling details persist. It is clear, however, that oxygen-regulated transcription of the *puf* operon of *R. capsulatus* initiates at a site upstream of the *pufQ* gene, and the promoter (P*pufl*) responsible has been sequenced (Bauer et al. 1988).

The structure of the oxygen-regulated promoter, P*pufl*, shows two features of interest. There are base sequences similar, but not identical, to the -12 and -24 consensus sequences of the so-called *ntrA* promoters of other bacteria (Gussin et al. 1986). *NtrA* promoters are recognized as transcription start sites in enteric bacteria by an RNA polymerase that carries a minor sigma factor, $\sigma 54$, in place of the predominant $\sigma 70$. The RNA polymerase(s) of *Rhodobacter* are not well enough studied to know whether analogous systems of promoters and polymerases exist in that genus, but the pattern does seem to be widespread among other bacteria. If the analogy does hold, it is interesting to note that *ntrA* promoters typically require a positive control protein, binding to the promoter via sequences adjacent to the consensus sequence (Kustu et al. 1989). The second feature of note in the P*pufl* sequence is an inverted repeat adjacent to the putative consensus sequence. This repeat is conserved in the region upstream from the *puh* operon, which is also oxygen-regulated, and Bauer et al. have suggested that this inverted repeat is a candidate binding site for a protein involved in the control of transcription by oxygen.

### 2.2 New Results

We have recently succeeded in isolating a protein from *R. capsulatus* that binds specifically to the inverted repeat of P*pufl*. We have purified this protein, which we call the puf promoter binding protein or PPBP, to near homogeneity by using a mobility-shift assay. As isolated from anaerobically grown cells, one molecule of PPBP binds preferentially to the operon-proximal half of the inverted repeat (IR-R in Bauer et al. 1988). Since the inverted repeat does not show perfect twofold symmetry, this preferecne may be understood on the basis of the sequence alone. When as-isolated PPBP is treated with alkaline phosphatase, its binding characteristics change so that two molecules bind per molecule of promoter. The phosphatase-treated PPBP can be phosphorylated by ATP in the presence of cell membranes, after which the mole ratio returns to 1:1. This entire pattern is consistent with the existence of a "two-component" regulatory system underlying oxygen control of transcription.

Two-component regulatory systems are widespread in the bacterial world, mediating cellular responses to such diverse stimuli as nitrogen status, phosphorus status, osmotic pressure, and chemotactic chemicals (reviewed by Stock et al. 1989). The two components referred to are a histidine protein kinase (HPK) and a response regulator (RR). The HPK is typically a membrane-bound protein that senses a signal and transduces the signal to the cellular machinery by transferring a phosphate group from ATP to an aspartate residue on the RR. The transfer involves a phosphohistidine residue on the sensory protein, hence the name HPK. The RR is frequently a DNA-binding transcriptional control factor. HPKs and RRs each have characteristic homologous domains in their structures which define the two families of proteins. Sometimes the homology is sufficient to allow cross-talk between regulatory systems.

M. Sganga and C. Bauer, working at Indiana University, have recently identified two genes, *regA* and *senC*, which are required for high level expression of the *puf* operon (Sganga and Bauer 1989 and unpublished results). Sequence analysis of these genes shows homology to the RR and HPK proteins of two-component systems.

M. Narro, C. Adams, and S. Cohen, working at Stanford University, have shown that mutations in IR-L of the *puf* promoter inverted repeat cause a constitutive phenotype, wherein transcription from the *puf* operon becomes much more resistant to oxygen (in press).

## 2.3 A New Model

Considering the results mentioned above, the model shown in Fig. 1 is proposed for the regulation of transcription of *puf* and related operons by oxygen. In the presence of sufficient oxygen, PPBP is dephosphorylated and binds to both parts

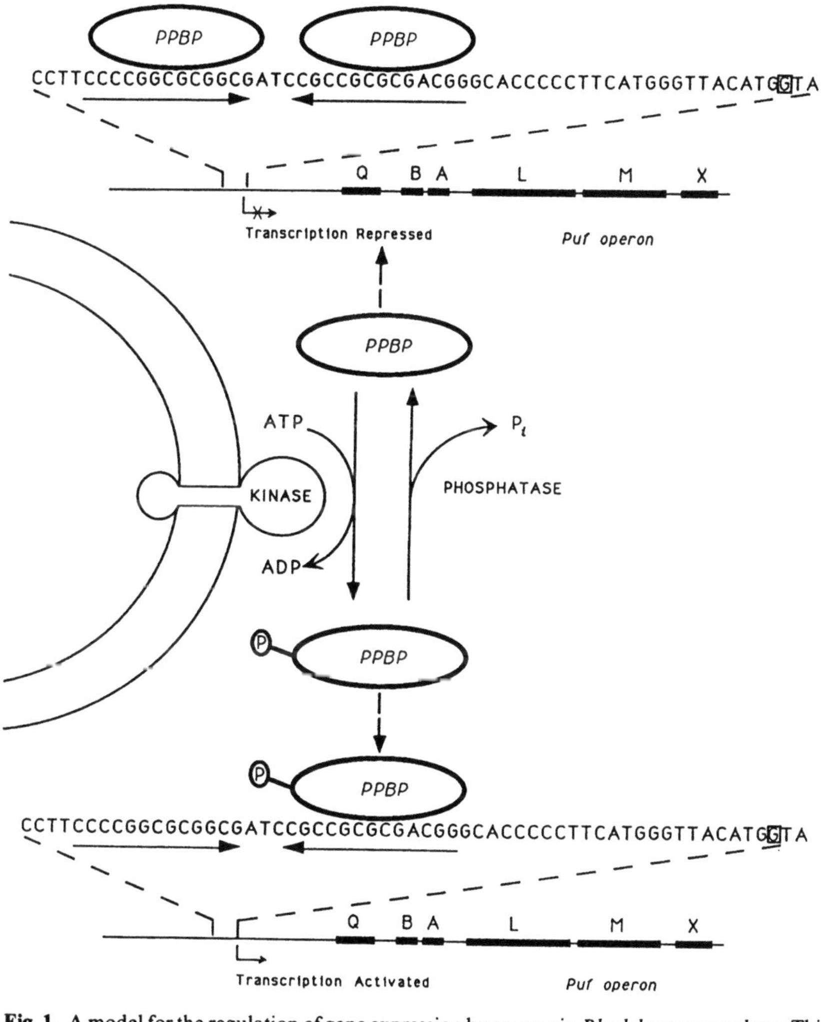


**Fig. 1.** A model for the regulation of gene expression by oxygen in *Rhodobacter capsulatus*. This model suggests two states for the *puf* promoter binding protein (PPBP): phosphorylated and dephosphorylated. In the phosphorylated state, PPBP activates transcription from the strong, oxygen-regulated promoter of the *puf* operon, $P_{pufP}$, and perhaps from other promoters in the same regulon (*puh, puc,* etc?). Activation is accompanied by the binding of PPBP to IR-R. Phosphorylation of PPBP is accomplished by a membrane-bound sensory kinase in the presence of ATP. The PPBP may be dephosphorylated by a cellular phosphatase. In its dephosphorylated state, PPBP acts as a repressor of *puf* transcription, binding to both IR-R and IR-L. The activities of the kinase or the phosphatase or both are thought to be regulated in response to oxygen tension, lower tension favoring the kinase; higher tension, the phosphatase

of the inverted repeat, repressing transcription. When oxygen tension is low, PPBP is phosphorylated, and in this form only IR-R is occupied, and transcription is activated. This model suggests that the membrane-bound kinase is a likely candidate for the oxygen sensor in the system, although its sensing of oxygen may well be indirect. The kinase might be activated, for example, by dihydroquinone in the quinone pool of cell membranes, or it might be inactivated by the oxidized form of quinone, or by oxygen itself. The phosphatase activity may reside in the same protein as the kinase, toggling between activities depending on the oxygen status of the cell. The gene called *senC* by Sganga and Bauer might encode this kinase. Sganga and Bauer's *regA* might code for PPBP, and we should have the sequence data to test that notion shortly.

## 3 Summary and Outlook

The results of the last few months have been quite exciting, because they suggest that we are closing in on a truly molecular understanding of how oxygen regulates gene expression. If the sensory protein is indeed a histidine protein kinase, we should be able to pin down what chemical entity affects its activity, especially because so much is known about the membranes of photosynthetic bacteria. One obvious candidate is the ubiquinone pool, since it would be expected to transduce oxygen tension changes into changes in redox state (quinone vs dihydroquinone), but this would leave unexplained the lack of effect on regulation observed in terminal oxidase-negative mutants. Finally, it should be noted that the kinase itself could react directly with oxygen.

## References


Bauer, C.E., Young, D.A. & Marrs, B.L. (1988) Analysis of the *Rhodobacter capsulatus puf* operon. Location of the oxygen-regulated promoter region and the identification of an additional *puf*-encoded gene. J. Biol. Chem. 263:4820–4827

Belasco, J.G., Beatty, J.T., Adams, C.W., von Gabain, A. & Cohen, S.N. (1985) Differential expression of photosynthesis genes in *R. capsulata* results from segmental differences in stability within the polycistronic *rxcA* transcript. Cell 40:171–181

Clark, W.G., Davidson, E. & Marrs, B. (1984) Variation of Levels of mRNA coding for antenna and reaction center polypeptides in *Rhodopseudomonas capsulata* in response to changes in oxygen concentration. J. Bacteriol. 157:945–948

Cohen-Bazire, G., Sistrom, W.R. & Stanier, R.Y. (1957) Kinetic studies of pigment synthesis by non-sulfur purple bacteria. J. Cell. Comp. Physiol. 49:25–68

Gussin, C.D., Ronson, C.W. & Ausubel, F.M. (1986) Regulation of nitrogen fixation genes. Ann. Rev. Genet. 20:567–591

Kustu, S., Santero, E., Keener, J., Popham, D. & Weiss, D. (1989) Expression of $\sigma$54 (*ntrA*)-dependent genes is probably united by a common mechanism. Microbiol. Rev. 53:367–376

Marrs, B. & Gest, H. (1973) Regulation of bacteriochlorophyll synthesis by oxygen in respiratory mutants of *Rhodopseudomonas capsulata*. J. Bacteriol. 114:1052–1057

Marrs, B., Cohen, S. & Taylor, D. (1981) Genetic engineering in a photosynthetic bacterium. In: Akoyunoglou, G. (ed) Photosynthesis V. Chloroplast development. Balaban International Science Services, Philadelphia, pp 687–694

Sganga, M. & Bauer, C. (1989) Cloning and characterization of the genes responsible for regulation of the photosynthetic apparatus by oxygen in *Rhodobacter capsulatus*. Abstr. Ann. Meet. Am. Soc. Microbiol. 1989:170

Schön, G. & Drews, G. (1968) Der Redoxzustand des NAD(P) und der Cytochrome b und $c_2$ in Abhängigkeit vom $pO_2$ bein einigen *Athiorhodaceae*. Arch.Mikrobiol. 61:317-326

Scolnik, P.A. & Marrs, B.L. (1987) Genetic research with photosynthetic bacteria. Ann. Rev. Microbiol. 41:703-726

Stock, J.B. Ninfa, A.J. & Stock, A.M. (1989) Protein phosphorylation and regulation of adaptive responses in bacteria. Microbiol. Rev. 53:450-490

Storz, G., Tartaglia, L.A. & Ames, B.N. (1990) Transcriptional regulator of oxidative stress-inducible genes: direct activation by oxidation. Science 248:189-194

Taylor, D.P., Cohen S.N., Clark, W.G. & Marrs, B.L. (1983) Alignment of genetic and restriction maps of the photosynthesis region of the *Rhodopseudomonas capsulata* chromosome by a conjugation-mediated marker rescue technique. J. Bacteriol. 154:580-590

van Niel, C.B. (1944) The culture, general physiology, morphology and classification of the nonsulfur purple and brown bacteria. Bacteriol. Rev. 8:1-118

# Phosphorus Assimilation and Its Control of Gene Expression in *Escherichia coli*

B.L. WANNER[1]

## 1 Introduction

Cells require usable forms of the elements C, N, O, P, and S to serve as major building blocks for innumerable biomolecules. Though C, N, O, and S compounds are both oxidized and reduced in various bacteria, P assimilation is relatively simple because P is used primarily only in the +5 valence state found naturally in phosphates and phosphonates. Inorganic orthophosphate (Pi) is the preferred P source and represses the transcription of phosphate (PHO) regulon genes for use of alternate forms of P. No control exists in which a specific P compound is needed for induction of a gene(s) required for utilization of that compound.

P is an essential component in membrane lipids, complex carbohydrates such as lipopolysaccharides, and nucleic acids. P plays a central role in energy metabolism and P is incorporated into many proteins posttranslationally. Therefore, Pi levels probably also regulate many genes important in cell physiology which are not directly related to P assimilation.

Over the five years since Pi control of gene expression was last reviewed (Wanner 1987b), our understanding has increased greatly, especially regarding P assimilation in the form of phosphonates, the molecular control of the PHO regulon, and the identification of Pi-regulated genes that are members of other regulatory systems. In this chapter I will briefly review our present understanding on assimilation of P in *E. coli* and the regulation of genes in response to changes in environmental Pi levels.

## 2 Results and Discussion

### 2.1 *Acquisition of Environmental Phosphates*

*E. coli* can use inorganic phosphates, organophosphate esters, or phosphonates. Pi enters the cell via the high Km Pi transport (Pit) or the low Km phosphate-specific transport (Pst) system. Pit is a single component transporter analogous to LacY and is expressed constitutively (Elvin et al. 1987). Pst is a periplasmic binding protein-dependent system similar to those for histidine, maltose, and ribose (Ames 1986). PstS is the periplasmic Pi-binding protein, PstA and PstC are integral membrane proteins, and PstB is similar to the HisP, MalK, and RbsB permeases which have conserved primary sequences for a nucleotide binding domain. PstB is

[1]Department of Biological Sciences, Purdue University, West Lafayette, IN 47907, USA

41. Colloquium Mosbach 1990
The Molecular Basis of Bacterial Metabolism

associated with the inner membrane via hydrostatic interactions, perhaps with PstA and PstC. In the absence of both Pit and Pst, cells no longer transport Pi and are organophosphate auxotrophs.

Glycerol-3-phosphate may enter cells for use as a P source via the binding protein-dependent UgpBAEC transporter, whose expression is Pi repressible (Overduin et al. 1988). In addition, GlpT and UhpT transport glycerol-3-phosphate and hexose-6-phosphates, respectively, although these permeases primarily transport the respective esters for use as C and energy sources (Ehrmann et al. 1987; Weston and Kadner 1988). Nontransportable organophosphates are also usable P sources. These organophosphate esters may enter the periplasm without the aid of a specific porin where, in Pi-limited cells, the esters are hydrolyzed by the nonspecific phosphomonoesterase bacterial alkaline phosphatase (Bap) which is the *phoA* gene product. The periplasmic acid phosphatase (AppA) or glucose-1-phosphatase (Agp) might hydrolyze specific esters which are nontransportable. AppA or Agp probably serve to provide cells with the C products of hydrolysis, as inferred from the mode of *appA* and *agp* gene regulation (Pradel et al. 1990). In any case, Pi released in the periplasm is actively transported via the Pit or Pst system.

Long chain polymetaphosphates (volutin; polyPi) may enter the periplasm through the PhoE porin (Benz et al. 1989), which is selective for anions, where polyPi is broken down by Bap. PhoE probably also allows entry of polyanions such as nucleic acids into the periplasm, which could then be degraded by the combined action of periplasmic nucleases and phosphatases, thereby providing cells with Pi together with a variety of usable C and N compounds.

### 2.2 *Phosphonate Utilization*

In addition to phosphates, *E. coli* can use a variety of phosphonates (Pn), which have direct carbon-to-phosphorus (C-P) bonds, as the sole P source. The use of a Pn with a substituted C-2, such as the natural 2-aminoethylphosphonate (AEPn), or an unsubstituted hydrocarbon group, such as the synthetic methylphosphonate (MPn), requires the complex *phn(psiD)* operon which probably encodes 14 open reading frames (Orfs) and may include a periplasmic Pn transport system together with a system for Pn biodegradation (Wackett et al. 1987; Chen et al. 1990; Wanner and Boline 1990).

The C-P bond is chemically very stable, therefore, its biodegradation may be difficult. Two pathways exist that act on different Pn types. The biodegradation of the C-2 substituted AEPn can proceed via a transamination-mediated dephosphonation in two steps: AEPn is first deaminated to phosphonacetaldehyde which is then hydrolyzed to Pi and acetaldehyde by a phosphonoacetaldehyde phosphohydrase (trivial name, phosphonatase). However, the phosphonatase pathway is not feasible for the biodegradation of an unsubstituted Pn, as the degradation products based on in vivo studies with alkyl- or benzyl-Pn are the corresponding hydrocarbons. Therefore, cleavage of the C-P bond in an unsubstituted Pn apparently proceeds via a direct dephosphonation, for which the trivial name C-P lyase was adopted. Since all *phn* mutations simultaneously abolish the use of either Pn type, both AEPn and MPn are probably metabolized by a C-P lyase in *E. coli*.

The complexity of the *phn* operon suggests that several *phn* gene products are likely to be involved in breaking the C-P bond.

The biochemical mechanism for C-P bond fission by a C-P lyase is poorly understood because it has been difficult to detect an activity in cell-free extracts. The first reports of cell-free cleavage of the C-P bond for an unsubstituted Pn appeared only recently. An *Enterobacter aerogenes* activity requires two components, E2 and E3, in which E2 is composed of six polypeptides of 14–24 kDa and E3 is an apparent homodimer of a 55-kDa polypeptide (Murata et al. 1989). A reductive mechanism for cleavage of the C-P bond has been proposed.

### 2.3 Assimilation of Pi into Organophosphate

Once Pi enters cells, it must be incorporated into the biophosphorus pool. Pi is probably assimilated directly into 1,3-diphosphoglycerate via substrate-level phosphorylation, acetylphosphate via mixed-acid fermentation, or ATP via oxidative phosphorylation (Fig. 1). There is no evidence for Pi regulation of the respective genes for glyceraldehyde-3-phosphate dehydrogenase (*gap*), phosphotransacetylase (*pta*), the $F_0F_1$ ATP synthase (*unc*), or other enzymes which use Pi as a substrate.

How the Pi moiety of an organophosphate is assimilated when transported via the UgpBAEC system is unclear. While glycerol-3-phosphate taken up by GlpT for use as a carbon source enters cells intact, glycerol-3-phosphate is unavailable as a carbon source when it is transported via the UgpBAEC transporter, which solely allows for its use as a P source. This observation has led to the suggestion that the UgpBAEC transporter might transfer the Pi moiety in glycerol-3-phosphate to another organophosphate, perhaps a phospholipid, which in turn serves as a phosphoryl donor.

Pi is the product of the *Enterobacter* Pn cleavage activity, though localization of this activity in the cell was not determined. It is also not known whether the *Enterobacter* and proposed *E. coli* C-P lyase activities are similar, because the organic products in the *Enterobacter* extract were not determined. Whether Pi or some phosphoryl compound is the direct in-vivo product of Pn metabolism is

Pi entry into organophosphate pool

a) Substrate level phosphorylation

i) Glycolysis, pentose phosphate shunt, Entner Douderoff Pathway

$$\text{Glyeraldehyde-3-Pi} + \text{Pi} \xrightarrow{gap} \text{1,3-diphosphoglyerate}$$

ii) Mixed acid fermentation

$$\text{Acetyl-CoA} + \text{Pi} \xrightarrow{pta} \text{Acetylphosphate} + \text{CoA}$$

b) Oxidative Phosphorylation

$$\text{ADP} + \text{Pi} \xrightarrow[\text{PMF}]{unc} \text{ATP}$$

**Fig. 1.** Assimilation of Pi into organophosphate

therefore not known. If fission of the C-P bond by a lyase involves a reductive cleavage mechanism, the P product could be in a reduced form, which might require an oxidation step by yet unidentified gene products.

### 2.4 Pi-Controlled Genes in E. coli

Studies with *lacZ* transcriptional fusions showed that more than 20 promoters are Pi-regulated in *E. coli* (Wanner 1987a,b). Consistent with these data was the observation that the synthesis of 81 proteins separated on two-dimensional polyacrylamide gels is markedly enhanced when *E. coli* growth is Pi-limited. In brief, 55 independent mutants made with Mu dI(*lacZ bla*) which showed a phosphate-starvation-inducible (*psi*) $Lac^+$ phenotype were characterized in a variety of ways, in order to identify the corresponding *psi* genes and to study both their genetic and physiological regulation. While it was anticipated that many *psi* genes would affect P assimilation, some *psi* genes were uncovered which had no direct or apparent involvement in P metabolism, as discussed below.

Only 15 of the 55 *psi* fusions had a phenotype expected for a mutant blocked in the assimilation of environmental P: 8 lie in the *phoA* gene; 3 lie in the *phn* operon; and 4 lie in or near the *ugpBAEC* operon. The *psiP*(*himA*)::*lacZ*(Mu dI) mutant, which was not anticipated, was identified because it failed to show the expected temperature-sensitive growth phenotype of a Mu dI mutant. A $HimA^-$, integration host factor ($Ihf^-$) mutant abolishes the temperature sensitivity due to an effect of the *himA* gene on Mu growth.

The sites of *psi*::*lacZ*(Mu dI)s in the *phoA*, *phn*(*psiD*), and *ugpBAEC* genes or operons were recently established by direct DNA sequence analysis of the fusion junctions (Fig. 2; Metcalf et al. 1990). DNA database searches led to the identification of eight additional *psi* fusions. The *psiF* fusion lies in an Orf downstream of the *phoA* gene, thus suggesting that the *phoA* gene is the promoter proximal gene of an operon. The *psi-51* fusion lies in the *glpABC* operon for the anaerobic glycerol-3-phosphate dehydrogenase. Three *psiQ* fusions lie in the *gltBDF* operon for glutamate synthase (GOGAT). Two *psiE* fusions lie in a sequenced Orf of yet unknown function near the *xylE* gene (Fig. 3). Our sequence information on other fusions should allow us to identify additional *psi* genes as more *E. coli* DNA sequences become available.

Overall, the *psi* genes may belong to two or more classes (Fig. 3). One class belongs to the PHO regulon which includes five genes or operons for acquisition of environmental P sources, in addition to the *phoBR* operon which encodes the PHO regulatory proteins PhoB and PhoR. Sixteen *psi* fusions are in the *phoA*-*psiF*, the *phn*(*psiD*), and the *ugpBAEC* operons. No *phoE* fusion was found; however, this is understandable since the number of fusions tested was small with respect to the number of genes involved, as shown by the fact that many *psi* genes were represented by only a single fusion. Also, no *pst* fusion was found even though the *pstSCAB*-*phoU* operon is highly regulated by Pi. In this case, none was expected because *pstSCAB*-*phoU* mutants abolish Pi repression; therefore, none would have shown a Pi-regulated Lac phenotype. Also, no *phoBR* fusion was anticipated because such a fusion abolishes autogenous control of the *phoBR* operon.

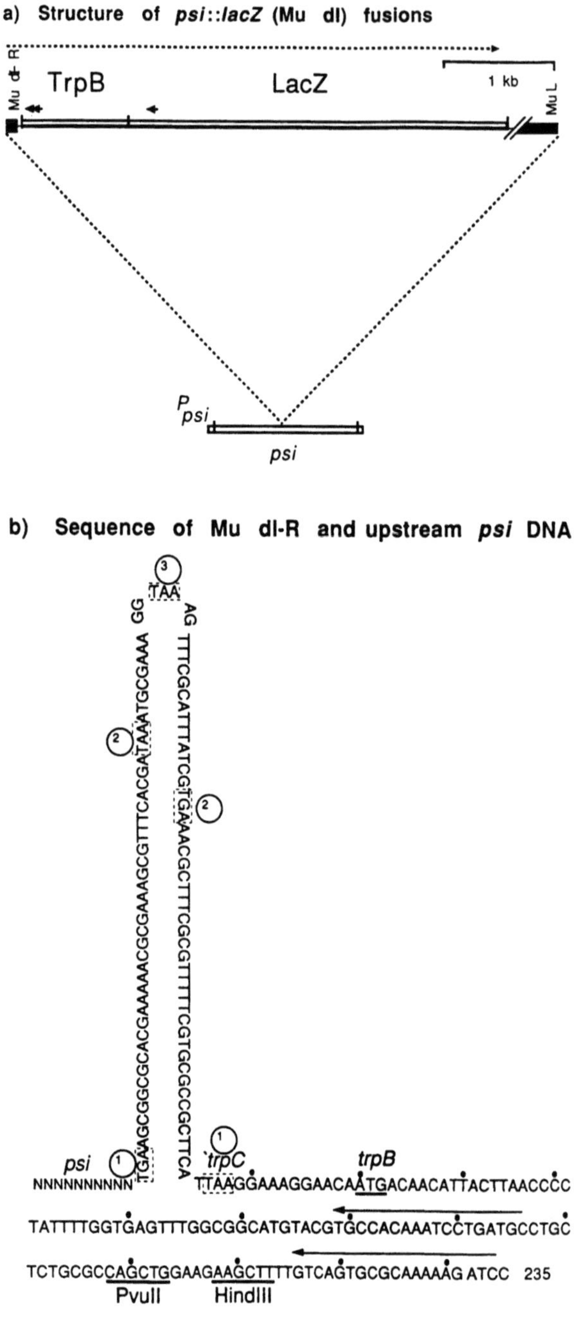


**Fig. 2.** **a** Structure of *psi*::*lacZ*(Mu dI) transcriptional fusions. **b** DNA sequence of the Mu dI-R fusion junctions. The DNA sequence information is from Metcalf et al. (1990)

a) PHO Regulon Genes

| Gene/operon | Map | Function |
|---|---|---|
| *phoA* (*psiA*), *psiF* (Orf-106) | 8.8' | Bap, Unknown |
| *phoE* | 5.8' | Polyanion Porin |
| *psiD* (*phnCDEFGHIJKLMNOP*) | 92.8' | Phosphonate Utilization |
| *pstSCAB* -*phoU* | 83.7' | Pi Transport and Repression |
| *ugpBAEC* (*psiB,C*) | 75.7' | Uptake *sn*-Glycerol-3-Phosphate |

b) *psi* Genes in Other Control Systems

| | | |
|---|---|---|
| *glpABC* (*psi* - *51*) | 48.7' | Glycerol-3-Phosphate Dehydrogenase (Anaerobic) |
| *gltBDF* (*psiQ*) | 69.4' | Glutamate Synthase (GOGAT) |
| *himA* (*psiP*) | 37.4' | Integration Host Factor |

c) Additional Pi-regulated Genes

| | | |
|---|---|---|
| *psiE* (Orf-136) | 91.4' | Unknown |
| > 10 other *psi* promoters | Various | Unknown |

**Fig. 3a-c.** Pi-regulated genes and operons in *E. coli* (see text)

A different class of *psi* genes includes members of other control systems. One *psi* fusion lies in the *glpABC* operon; three *psi* fusions lie in the *gltBDF* operon; and one lies in the *himA* gene (Fig. 3). No studies exist that show an effect of Pi on the expression of the anaerobic glycerol-3-phosphate dehydrogenase (GlpABC), glutamate synthase (GltBDF), or integration host factor (HimA). The apparent Pi control of these promoters could reflect direct effects due to Pi limitation, indirect physiological effects due to inactivation of the respective gene, or unknown effects due to the genetic background used to identify the *psi* fusions. For instance, Pi control of the *psiP*(*himA*)::*lacZ* fusion was abolished in $Aro^+$ transductants of the *aroB* mutant that was fortuitously used to isolate most of the *psi*::*lacZ* fusions (Wanner 1987b). Therefore the Pi effect on the *psiP*(*himA*) fusion may be mediated via an effect on aromatic amino acid synthesis which indirectly affects expression of the complex *pheST-himA* operon (Friedman 1988).

Pi control of the *glpABC*(*psi-51*) and *gltBDF*(*psiQ*) fusions is independent of the PHO regulon, as neither PhoB nor PhoR is required for their expression. But, there could still be a direct effect of Pi on the control of these, or other Pi-regulated *psi* promoters, via different regulators. There may exist separate Pi regulatory systems which respond to different Pi levels. The PHO regulon responds to extracellular Pi levels in the $\mu$M range, as expected for the regulation of genes concerned with an acquisition of environmental P sources. Since the cellular [Pi] is several mM, a different Pi regulatory system may operate, perhaps in the control of genes for enzymes in central metabolism, which responds to [Pi] changes in the mM Pi range. The *psi* genes that are members of other control systems could be subject to a general Pi control separate from PHO regulon control.

The synthesis of enzymes such as the anaerobic glycerol-3-phosphate dehydrogenase (GlpABC) which provide Pi for glyceraldehyde-3-phosphate dehydrogenase (Gap) might be expected to be Pi-regulated. In Pi-starved cells, GlpABC could lead to conversion of endogenous glycerol-3-phosphate to dihydroxyacetone phosphate (DHAP) and thereby act to salvage Pi for glycolysis. Pi salvage would occur if excess DHAP is converted to methylglyoxal and Pi. Accordingly, other glycolysis genes such as a gene for methylglyoxal synthetase may also be Pi-regulated.

While I have suggested why a *glpABC* fusion may have been identified as a *psi* gene, our finding of *gltBDF* fusions could be for a different reason. Both the nitrogen regulator (NtrC) and PHO regulator (PhoB) are activated by protein phosphorylation (Stock et al. 1989). There might exist cross talk among regulators and sensory protein kinases that share sequence and, presumably, structural features in common. Therefore the apparent Pi control of the *gltBDF* operon might involve a form of cross talk between the PHO and Ntr systems which is physiologically relevant.

### 2.5 PHO Regulon Control

Pi repression of the PHO regulon requires the high affinity PstSCAB transporter and PhoU. Pi control is independent of Pit (Fig. 4). Derepression due to Pi limitation occurs only when the environmental [Pi] falls below about 4 μM, i.e., an amount of Pi necessary to saturate PstS ($K_d$ = 0.8 μM). No concomitant decrease in the cellular [Pi] occurs in the time before derepression of the PHO regulon. Therefore, a transmembrane signal must emanate from PstSCAB which regulates the PHO regulon. PstS is involved in Pi signal transduction because numerous *pstS* alleles, including missense mutations, abolish Pi repression. There are no definitive data either way for a direct involvement of PstC or PstA in Pi repression. Indeed, the processes of Pi transport and repression can be uncoupled, since missense changes in PstC abolish Pi transport without affecting Pi repression (Cox et al. 1989). These data corroborate earlier evidence that Pi transport per se is not responsible for Pi repression. A *pstA* nonsense mutation leads to low constitutive

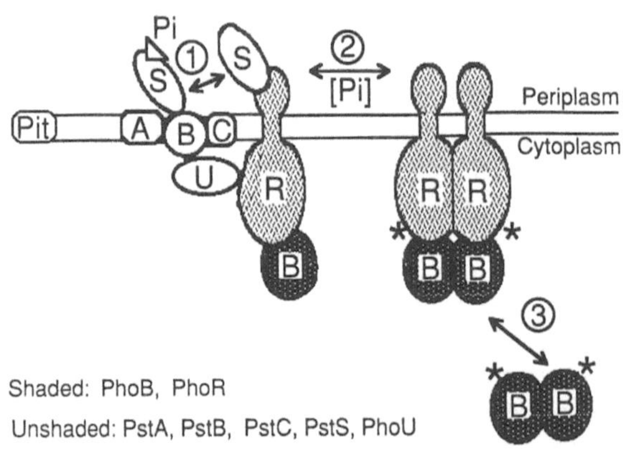


**Fig. 4.** Model for Pi control of the PHO regulon (see text)

expression of the PHO regulon (Cox et al. 1988); its effect could be due to polarity on the downstream *pstB* and *phoU* genes (Fig. 3). As no nonpolar *pstA* mutants with a Pi phenotype are available, its role in Pi repression is uncertain. PstB is involved in both Pi transport and repression because site-directed changes in the PstB nucleotide-binding consensus domain simultaneously abolish both processes (Cox et al. 1989). PhoU, which is the distal product of the *pstSCAB-phoU* operon, may act only in repression, because a *phoU* mutation abolishes Pi repression without affecting transport (Surin et al. 1986). Unfortunately, all studies on PhoU concern a single allele, *phoU35*, which causes a missense change (G. Cox, pers. commun.). PhoU may act in a way similar to $P_{II}$, the *glnD* gene product in the Ntr system, which is required for the regulated dephosphorylation of NtrC-Pi (Magasanik 1988). Whatever the role for PhoU is, there is an apparent need for more PhoU during Pi limitation, since derepression of the PHO regulon turns on the *pstSCAB-phoU* operon. Perhaps more PhoU is needed to reestablish repression following a period of Pi limitation than that which is needed to maintain a normal state of repression.

PhoB is the transcriptional activator, and PhoR is the Pi sensor which mediates a conversion of PhoB between active and inactive forms. The model in Fig. 4 depicts how PstSCAB, PhoU, PhoR, and PhoB may interact in Pi transport and repression. Accordingly, when the environmental [Pi] exceeds about 4 $\mu$M, the PHO regulon, including the PstSCAB transporter, is repressed and Pi is primarily transported via the Pit permease. Under these conditions, PstS is probably saturated with Pi, and Pi-bound PstS probably interacts with PstB. [A direct interaction between PstS and PstB is inferred by analogy with HisJ and HisP, respectively (Prossnitz et al. 1989).] The small amount of Pi transported by the repressed level of the PstSCAB transporter would be insufficient to support growth. An interaction of PhoU with both PstB and PhoR could allow PhoU to maintain repression when PstSCAB is in low amounts and saturated with Pi, and to reestablish repression when PstSCAB and PhoR are in high amounts and there is an excess of Pi. The relative amounts of PstSCAB, PhoU, and PhoR under these two conditions would determine whether PhoU is primarily associated with PstB or PhoR.

The finding that PhoB and PhoR (Makino et al. 1986a,b) share sequence similarities at the protein level to regulator and sensor members of two-component regulatory gene systems has provided new insights into how the PHO regulon is regulated at the biochemical level. Transcriptional activation in two-component regulatory systems involves autophosphorylation of the sensor protein, and protein phosphorylation and dephosphorylation of the regulator by its partner sensor protein (Bourret et al. 1989). By analogy, PhoR was recently shown to act as an autophosphorylase and protein kinase for PhoB. Thus, PhoR probably converts PhoB between its active and inactive forms by phosphorylating and dephosphorylating PhoB (Makino et al. 1989).

Pi control may involve three stages (Fig. 4): (1) a decrease in [Pi] may free PstS to interact with a PhoR domain in the periplasm. The PstS-PhoR interaction may be analogous to how certain periplasmic binding proteins interact with membrane transducers in bacterial chemotaxis (Burrows et al. 1989). PstS may interact with PhoR in a way that causes PhoR to dissociate from PhoU in the cytoplasm. This dissociation may promote PhoR dimer formation thus converting PhoR from its repressor form into its activator form, which were formerly denoted as $PhoR^R$ and

PhoR$^A$, respectively (Wanner 1987b). (2) The PhoR dimers may no longer interact with PstS and the dimeric, activator form PhoR* may act as an autophosphorylase. (3) PhoR* may phosphorylate PhoB and thereby convert inactive PhoB into its active form PhoB*. As long as [Pi] is insufficient to saturate the PstSCAB transporter, PhoU may remain associated with PstB. When Pi saturates PstB, PhoU may be released to interact with PhoR. An increased amount of PhoU may be needed to convert PhoR into its repressor form which may dephosphorylate and thereby inactivate PhoB.

## *2.6 The creABCD (phoM) Operon, Cross Talk, and Competition Among Two-Component Regulators*

Earlier studies suggested that there was a dual control over the PHO regulon. Pi control required PhoB and PhoR together with PstSCAB and PhoU. A separate control, which was evident only in *phoR* mutants, still required PhoB but was independent of PstSCAB and PhoU. This alternate control was quite substantial: it accounted for a constitutive level of PHO regulon expression that was about 30% of the transcription during Pi limitation of PhoR$^+$ cells. Furthermore, a gene called *phoM* was needed for this transcription in *phoR* mutants. (The *phoM* nomenclature was adopted because the *phoM* phenotype was masked in PhoR$^+$ cells.) It was thought that an unknown trigger could induce *phoM*-dependent transcription of the PHO regulon in *phoR*$^+$ cells (Wanner 1987b). The *phoM* nomenclature is now inappropriate, as this gene has no apparent role in PHO regulon control. We recently renamed the operon that includes the *phoM* gene as the catabolite regulator (*cre*) operon, due to its involvement in regulating genes in response to environmental catabolites (Wanner et al. 1988ab; in prep.).

The *cre* operon encodes four products: CreA, CreB, CreC, and CreD (Fig. 5), as predicted from the DNA sequence of the *creABCD* (formerly *phoM*) operon

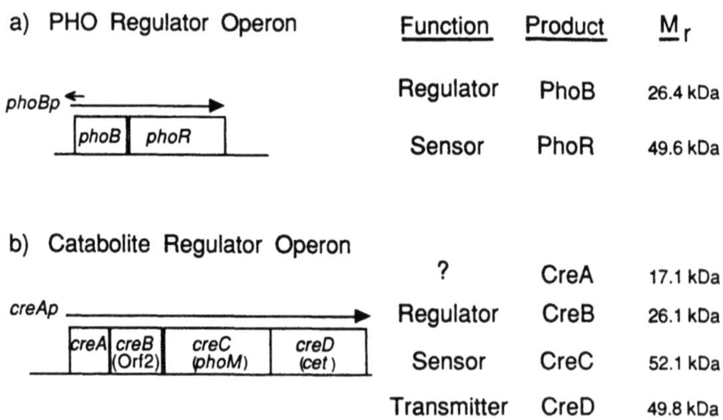


**Fig. 5a,b.** The *phoBR* and *creABCD* (*phoM*) regulator operons in *E. coli*. *Arrows* show transcripts determined with *lacZ* fusions in this laboratory which include an antisense RNA for the *phoBR* operon (Wanner and Chang 1987; in prep.). The transmitter function for CreD is according to Kofoid and Parkinson (1988). The *creD* gene is the same as *cet* (Drury and Buxton 1988)

(Amemura et al. 1986; Drury and Buxton 1988). CreB and CreC probably act as a regulator and sensor, respectively, in response to catabolites in a manner analogous to the action of PhoB and PhoR or similar partner proteins in other two-component regulatory systems. The striking protein sequence similarity between PhoR and CreC(PhoM) is probably responsible for cross talk in which CreC can activate, i.e., phosphorylate, PhoB. Cross talk is apparently seen only in *phoR* mutants, because in *phoR*$^+$ cells, PhoR would specifically dephosphorylate PhoB when Pi is in excess.

Our understanding that CreC(PhoM)-dependent transcription has no normal role in the control of the PHO regulon came from our recent studies on *creABCD*-dependent gene regulation. First, the intermediate (30%) constitutive level of PHO regulon expression in *phoR* mutants is actually due to a mutant form of the *creABCD* (*phoM*) operon in which the *creB* gene is mutated. The *creB510(pho-150)* mutation that is responsible for constitutivity of the PHO regulon in *phoR* mutants fortuitously exists in many laboratory strains of *E. coli* K-12 (Wanner 1987c). Second, PHO regulon expression in *phoR* mutants that are wild-type, i.e., *cre(ABCD)*$^+$, is highly regulated by several environmental factors, including catabolites. Both the CreC(PhoM)-dependent constitutivity of the PHO regulon in *creB* mutants and the catabolite-regulated expression of the PHO regulon in *creB*$^+$ cells can be explained by a competition between CreB and PhoB for the CreC (PhoM) protein kinase in a *phoR* mutant. It is this competition which is apparently also responsible for the metastable expression (clonal variation) of the PHO regulon in *phoR* mutants (Wanner 1986; Wanner et al. 1988a). Since competition is abolished when the *creB* gene is mutated, *creB phoR* double mutants express the PHO regulon constitutively in a manner that requires CreC(PhoM). The CreC-dependent catabolite control of the PHO regulon in *creB*$^+$ *phoR* mutants is due in part to regulation of the *creABCD* operon, for such catabolites decrease transcription at the *creA* promoter (in prep.). In addition, induction of the PHO regulon turns on the *phoBR* operon and the autogenous control of the *phoBR* operon leads to increased PhoB synthesis which in turn competes more effectively with CreB. As a consequence, conditions that induce the PHO regulon in *phoR* mutants lead to an induced state which is heritable via a mechanism that involves protein phosphorylation and protein-protein interactions (Wanner et al. 1988a).

## 3 Summary and Outlook

Recent studies on the *phn(psiD)* mutants were instrumental in showing Pi control of a C-P bond cleavage activity, which has yet to be detected in *E. coli* cell-free extracts. New molecular studies on the complex *phn(psiD)* operon should prove useful in understanding Pn biodegradation.

Studies on other *psi* genes suggest that there may exist Pi-regulated genes in *E. coli* which are not directly concerned with the acquisition of environmental P. There may exist a general Pi control over genes that are members of other regulatory systems.

Genetic and molecular studies on the PHO regulon showed that Pi signal transduction is fundamentally similar to nitrogen control, bacterial chemotaxis, and osmoregulation, in that each involves autophosphorylation, protein phosphorylation, and dephosphorylation of a regulator by a partner sensor protein. The PHO regulon may be unique among two-component regulatory systems in that a periplasmic binding protein-dependent transporter is involved in Pi signal transduction.

The *creABCD(phoM)* operon has no normal role in the control of the PHO regulon. Rather, it regulates other genes in response to environmental catabolites. Though cross talk is almost certainly the molecular basis for how CreC(PhoM) acts in place of PhoR in *phoR* mutants, it is unclear whether or not cross talk has a normal physiological role in bacteria.

*Acknowledgment.* I thank B. Chang, J. Cui, W. Metcalf, P. Steed, and M. Wilmes-Riesenberg for recent contributions to studies in my laboratory and for comments on the manuscript and J. Sheridan for secretarial assistance. This work was supported by GM35392 from NIH and DMB8804977 from NSF.

## References

Amemura, M., Makino, K., Shinagawa, H. & Nakata, A. (1986) Nucleotide sequence of the *phoM* region of *Escherichia coli*: four open reading frames may constitute an operon. J. Bacteriol. 168:294–302

Ames, G.F.-L. (1986) Bacterial periplasmic transport systems: structure, mechanism, and evolution. Annu. Rev. Biochem. 55:397–425

Benz, R., Schmid, A., Van der Ley, P. & Tommassen, J. (1989) Molecular basis of porin selectivity: membrane experiments with OmpC-PhoE and OmpF-PhoE hybrid proteins of *Escherichia coli* K-12. Biochim. Biophys. Acta 981:8–14

Bourret, R.B., Hess, J.F., Borkovich, K.A., Pakula, A.A. & Simon, M.I. (1989) Protein phosphorylation in chemotaxis and two-component regulatory systems of bacteria. J. Biol. Chem. 264:7085–7088

Burrows, G.G., Newcomer, M.E. & Hazelbauer, G.L. (1989) Purification of receptor protein Trg by exploiting a property common to chemotactic transducers of *Escherichia coli*. J. Biol. Chem. 264:17309–17315

Chen, C.-M., Ye, Q., Zhu, Z., Wanner, B.L. & Walsh, C.T. (1990) Molecular biology of carbon-phosphorus bond cleavage: cloning and sequencing of the *phn* (*psiD*) genes involved in alkylphosphonate uptake and C-P lyase activity in *Escherichia coli* B. J. Biol. Chem. 265:4461–4471

Cox, G.B., Webb, D., Godovac-Zimmermann, J. & Rosenberg, H. (1988) Arg-220 of the PstA protein is required for phosphate transport through the phosphate-specific transport system in *Escherichia coli* but not for alkaline phosphatase repression. J. Bacteriol. 170:2283–2286

Cox, G.B., Webb, D. & Rosenberg, H. (1989) Specific amino acid residues in both the PstB and PstC proteins are required for phosphate transport by the *Escherichia coli* Pst system. J. Bacteriol. 171:1531–1534

Drury, L.S. & Buxton, R.S. (1988) Identification and sequencing of the *Escherichia coli cet* gene which codes for an inner membrane protein, mutation of which causes tolerance to colicin E2. Mol. Microbiol. 2:109–119

Ehrmann, M., Boos, W., Ormseth, E., Schweizer, H. & Larson, T. (1987) Divergent transcription of the *sn*-glycerol-3-phosphate active transport (*glpT*) and anaerobic *sn*-glycerol-3-phosphate dehydrogenase (*glpA glpC glpB*) genes of *Escherichia coli* K-12. J. Bacteriol. 169:526–532

Elvin, C.M., Hardy, C.M. & Rosenberg, H. (1987) Molecular studies on the phosphate (inorganic) transport (*pit*) gene of *Escherichia coli*: identification of the *pit*$^+$ gene product and physical mapping of the *pit-gor* region of the chromosome. Mol. Gen. Genet. 204:477–484

Friedman, D.I. (1988) Integration host factor: a protein for all reasons. Cell 55:545–554

Kofoid, E.C. & Parkinson, J.S. (1988) Transmitter and receiver modules in bacterial signaling proteins. Proc. Natl. Acad. Sci. USA 85:4981–4985

Magasanik, B. (1988) Reversible phosphorylation of an enhancer binding protein regulates the transcription of bacterial nitrogen utilization genes. TIBS 13:475–479

Makino, K., Shinagawa, H., Amemura, M. & Nakata, A. (1986a) Nucleotide sequence of the *phoB* gene, the positive regulatory gene for the phosphate regulon of *Escherichia coli* K-12. J. Mol. Biol. 190:37–44

Makino, K., Shinagawa, H., Amemura, M. & Nakata, A. (1986b) Nucleotide sequence of the *phoR* gene, a regulatory gene for the phosphate regulon of *Escherichia coli*. J. Mol. Biol. 192:549–556

Makino, K., Shinagawa, H., Amemura, M., Kawamoto, T., Yamada, M. & Nakata, A. (1989) Signal transduction in the phosphate regulon of *Escherichia coli* involves phosphotransfer between PhoR and PhoB proteins. J. Mol. Biol. 210:551–559

Metcalf, W.W., Steed, P.M. & Wanner, B.L. (1990) Identification of phosphate-starvation-inducible genes in *Escherichia coli* K-12 by DNA sequence analysis of *psi*::*lacZ*(Mu dI) transcriptional fusions. J. Bacteriol. 172:3191–3200

Murata, K., Higaki, N. & Kimura, A. (1989) A microbial carbon-phosphorus bond cleavage enzyme requires two protein components for activity. J. Bacteriol. 171:4504–4506

Overduin, P., Boos, W. & Tommassen, J. (1988) Nucleotide sequence of the *ugp* genes of *Escherichia coli* K-12: homology to the maltose system. Mol. Microbiol. 2:767–775

Pradel, E., Marck, C. & Boquet, P.L. (1990) Nucleotide sequence and transcriptional analysis of the *Escherichia coli agp* gene encoding periplasmic acid glucose-1-phosphatase. J. Bacteriol. 172:802–807

Prossnitz, E., Gee, A. & Ferro-Luzzi Ames, G. (1989) Reconstitution of the histidine periplasmic transport system in membrane vesicles. Energy coupling and interaction between the binding protein and the membrane complex. J. Biol. Chem 264:5006–5014

Stock, J.B., Ninfa, A.J. & Stock, A.M. (1989) Protein phosphorylation and regulation of adaptive responses in bacteria. Microbiol. Rev. 53:450–490

Surin, B.P., Dixon, N.E. & Rosenberg, H. (1986) Purification of the PhoU protein, a negative regulator of the *pho* regulon of *Escherichia coli* K-12. J. Bacteriol. 168:631–635

Wackett, L.P., Wanner, B.L., Venditti, C.P. & Walsh, C.T. (1987) Involvement of the phosphate regulon and the *psiD* locus in the carbon-phosphorus lyase activity of *Escherichia coli* K-12. J. Bacteriol. 169:1753–1756

Wanner B.L. (1986) Bacterial alkaline phosphatase clonal variation in some *Escherichia coli* K-12 *phoR* mutant strains. J. Bacteriol. 168:1366–1371

Wanner, B.L. (1987a) Molecular cloning of Mud (*bla lacZ*) transcriptional and translational fusions. J. Bacteriol. 169:2026–2030

Wanner, B.L. (1987b) Phosphate regulation of gene expression in *Escherichia coli*. In: Neidhardt, F.C., Ingraham, J., Low, K.B., Magasanik, B., Schaechter, M. & Umbarger, H.E. (eds) *Escherichia coli* and *Salmonella typhimurium* cellular and molecular biology, Vol. 2. Am. Soc. Microbiol., Washington D.C., pp. 1326–1333

Wanner, B.L. (1987c) Control of *phoR*-dependent bacterial alkaline phosphatase clonal variation by the *phoM* region. J. Bacteriol. 169:900–903

Wanner, B.L. & Boline, J.A. (1990) Mapping and molecular cloning of the *phn* (*psiD*) locus for phosphonate utilization in *Escherichia coli*. J. Bacteriol. 172:1186–1196

Wanner B.L. & Chang, B.-D. (1987) The *phoBR* operon in *Escherichia coli* K-12. J. Bacteriol. 169:5569–5574

Wanner, B.L., Wilmes, M.R. & Young, D.C. (1988a) Control of bacterial alkaline phosphatase synthesis and variation in an *Escherichia coli* K-12 *phoR* mutant by adenyl cyclase, the cyclic AMP receptor protein, and the *phoM* operon. J. Bacteriol. 170:1092–1102

Wanner, B.L., Wilmes, M.R. & Hunter, E. (1988b) Molecular cloning of the wild-type *phoM* operon in *Escherichia coli* K-12. J. Bacteriol. 170:279–288

Weston, L.A. & Kadner, R.J. (1988) Role of *uhp* genes in expression of the *Escherichia coli* sugar-phosphate transport system. J. Bacteriol. 170:3375–3383

# Regulation of Gene Expression by Iron

V. BRAUN, S. SCHÄFFER, K. HANTKE, and W. TRÖGER[1]

## 1 Introduction

Iron serves as a central regulatory substance which controls the expression of almost 40 genes in *Escherichia coli*. Why is this so? Iron, with the exception of some lactobacilli, is required by all organisms in rather high concentrations (between $10^5$ and $10^6$ ions per microbial cell) [1] since it is contained in the reaction centers of many redox enzymes in the cytoplasm and the cytoplasmic membrane. Depending on the protein environment, $Fe^{2+}/Fe^{3+}$ spans the unusually large standard redox potential range from +300 to -500 mV, which makes it an ideal redox cofactor. Despite the great abundance of iron in nature, iron supply poses great problems for organisms growing under aerobic conditions. $Fe^{3+}$ occurs at pH 7 as a hydroxyaquo polymer with a free $Fe^{3+}$ concentration in the order of $10^{-18}$ M ($10^3$ ions per ml). This is certainly not enough to supply $10^9$ bacteria per ml with $10^{14}$ $Fe^{3+}$ ions per generation. To fulfill their iron demand, bacteria developed elaborate iron supply systems which include iron complexing substances, called siderophores, and $Fe^{3+}$-siderophore transport systems. Synthesis of the siderophores and the transport systems are subject to iron control. The enzymes and transport proteins involved are synthesized under conditions of iron starvation. In addition, a number of bacterial toxins are formed at low iron supply. As far as these toxins damage eukaryotic cells, bacteria may gain access to intracellular iron stores, so that the toxins also contribute to the bacterial iron supply. However, there are iron-regulated genes with no apparent relation to the iron supply, such as those which determine luminescence of *Vibrio harveyi* [2], swarming of *Vibrio parahemolyticus* [3], manganese and the iron-dependent superoxide dismutases [4], a flavodoxin protein of *Anacystis nidulans* [5], nitrogenase (via the NifL repressor) of *Klebsiella pneumoniae* [6], and the S-pili of *E. coli* (J. Hacker, pers. commun.). In addition, *E. coli* mutants devoid of iron regulation are unable to grow on succinate, fumarate, and acetate [7]. In these cases, iron regulation seemingly serves other purposes than iron supply. Iron acts as an environmental signal which is used to adapt cells to a particular ambient. The low concentration of available iron is a typical parameter of this environment, which is recognized by the bacteria and transformed into gene expression. Interestingly, not iron as a substance but the lack of iron serves as a signal, which turns on global regulation of a number of genes, mostly organized in operons, and scattered throughout the chromosome and plasmids (Fig. 1).

[1] Mikrobiologie/Membranphysiologie and Organische Chemie, Universität Tübingen, W-7400 Tübingen, FRG

41. Colloquium Mosbach 1990
The Molecular Basis of Bacterial Metabolism

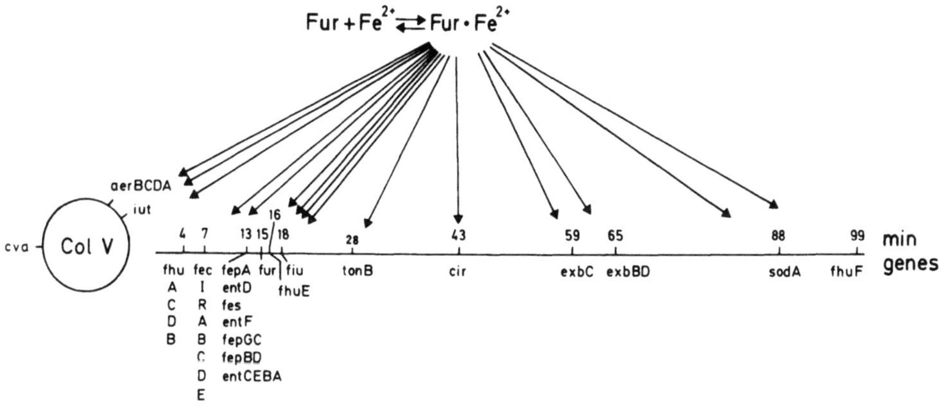


**Fig. 1.** Genes of *E. coli* known to be regulated by the Fur repressor. The genes have been ordered at their positions along the linearized 100 min linkage map of the *E. coli* chromosome. The sequence of the letters corresponds to the arrangement of the genes within the operons; for example *fhuA CDB* indicates the gene order *fhuA fhuC fhuD fhuB* transcribed in this direction. The ColV plasmid encodes the iron-regulated *cva* gene for colicin V (31), aerobactin biosynthesis (*aer*), and the outer membrane receptor *iut* for iron(III)-aerobactin transport

## 2 Iron Regulation via the Fur Repressor of *E. coli*

Siderophores are synthesized and released into growth media containing low concentrations of available iron [8]. Proteins can be related to $Fe^{3+}$-siderophore transport by the appearance of new bands on electropherograms loaded with membranes of cells grown at iron-limiting conditions. This has been seen with the outer membrane proteins of *E. coli* [9] and *Salmonella typhimurium* [10]. In contrast, the amount of iron transport proteins present in the cytoplasmic membrane is too low to be detected by staining, and no prominent radiolabeled bands appear under conditions of iron starvation. Therefore, regulation was studied at the level of gene expression by using a derivative of phage Mu [Mud1(Ap*lac*)] which carries the genes for lactose utilization but lacks the regulatory elements of the *lac* operon [11]. By integration of the phage into a gene in the direction of its transcription, the lactose structural genes come under the control of that gene. Phage Mu integrates spontaneously at random locations into the bacterial chromosome and plasmids. Mud1(Ap*lac*) serves as a convenient tool for measuring gene transcription rates. In the case of iron-regulated genes, synthesis of $\beta$-galactosidase is enhanced under low iron growth conditions. We have applied this technique to measure transcription of iron-regulated genes encoding outer and inner membrane transport proteins, and of enzymes catalyzing aerobactin (a siderophore) biosynthesis [12–17]. The transcription rates of the 14 genes studied differed, and were enhanced 3–27-fold at depleted, compared to repleted, iron conditions [18]. The Mud1(Ap*lac*) promoter fusions also provided a convenient way to isolate regulation-deficient mutants. After chemical mutagenesis of a strain carrying a Mud1(Ap*lac*) insertion in the *fhuA* gene (which determines an outer membrane receptor for ferrichrome transport), derivatives were selected which synthesized high amounts of β-galactosidase (deep red colonies on MacConkey plates containing sufficient amounts of iron). Those mutants which produced, irrespective of the iron supply, increased amounts of all (six) of the iron-regulated outer membrane proteins were mutated at a single locus, designated *fur* [12] which was mapped at minute 15.5 (now close to the *glnS* gene) of the *E. coli* linkage map [18]. The DNA of this locus was cloned and shown to encode a protein which repressed transcription of the iron-regulated genes in the presence of iron [19]. Sequencing of the gene [20] revealed an open reading frame that encoded a protein with a molecular weight of 16 795 daltons (148 amino acids) which agreed (within the accuracy of the method employed) with the size of the polypeptide determined by polyacrylamide gel electrophoresis (18 000 daltons) [20]. Release of the C-terminal Lys residue by carboxypeptidase B, and of Lys and Arg by trypsin from the internal Lys-Lys and Lys-Arg sequences, confirmed the amino acid sequence derived from the nucleotide sequence [20]. The outstanding property of the protein is the high content of histidine (12 residues).

The Fur protein was isolated on a preparative scale from an overexpressing clone constructed by the Monsanto Company [21]. A single metal ion affinity chromatography over zink iminodiacetate agarose resulted in an electrophoretically homogeneous protein. Using this sample, binding of the Fur protein to a DNA region upstream of the aerobactin synthesis genes was studied. This region was thought to contain the proposed operator region controlling aerobactin gene transcription. Depending on the presence of $Fe^{2+}$ (or $Mn^{2+}$ which is less prone to oxidation than $Fe^{2+}$) the Fur protein protected a 31-base-pair fragment from DNase I cleavage, and a second region of 19 base pairs at higher Fur concentrations [22]. Binding to the first site was enhanced about 1000-fold by the presence of $Mn^{2+}$. This site spans the promoter from seven base pairs upstream of the –35 region to the first base pair of the –10 region. $Mn^{2+}$-Fur also protected cleavage by the *Hinf*I restriction enzyme in the aerobactin promoter region. These results were supported by $Mn^{2+}$ and Fur-dependent inhibition of β-galactosidase synthesis in an in vitro transcription-translation system programed by an *iucC* (*aerC*)::*lacZ* operon fusion [23]. Binding of the Fur repressor in the presence of $Mn^{2+}$ was also shown to occur at two promoter regions upstream of the *cir* gene [24], which encodes an iron-regulated outer membrane protein involved in $Fe^{3+}$-siderophore uptake of the catecholate type. Moreover, a deletion analysis of the DNA region upstream of the structural gene (*sltA*) encoding the Shiga-like toxin of *E. coli* localized the iron-regulated promoter [25]. The consensus sequence for the Fur binding site derived from this study, was confirmed by introducing a synthetic nucleotide with the consensus sequence in front of an *ompF-lacZ* fusion, which subsequently acquired iron-regulated *lacZ* expression [25]. Sequences similar to those found upstream of *aerC, cir* and *sltA* were also found upstream of a number of iron-regulated genes, from which a consensus sequence for the Fur repressor binding region has been derived (Fig. 2). The in vivo results obtained with fusions between iron-regulated operons and the *lacZ* gene, and the data of in vitro protection experiments with isolated Fur protein, leave little doubt that the Fur protein is converted to a transcriptional repressor upon binding of $Fe^{2+}$.

| Gene | Nucleotide No.[a] | Sequence |
|---|---|---|
| in *E.coli* | | |
| *aerA* (*iucA*) | 349 | G A T A A T G A G A A T C A T T A T T |
| *aerA* (*iucA*) | 369 | C A T A A T T G T T A T T A T T T T A |
| *fhuA* | 383 | C T T T A T A A T A A T C A T T C T C |
| *fepA* | 415 | A T T A T T G A T A A C T A T T T G C |
| *fepA* | 427 | T A T A T T G A T A A T A T T A T T G |
| *fhuE* | 62 | T A C A A A C A A A A T T A T T C G C |
| *fhuE* | 40 | G C G T A T A T T T C T C A T T T G C |
| *fecA* | 144 | G A A A A T A A T T C T T A T T T C G |
| *fecA* | 138 | T G T A A G G A A A A T A A T T C T T |
| *cir* | 2 | T G G A T T G A T A A T T G T T A T C |
| *cir* | 8 | G A T A A T T G T T A T C G T T T G C |
| *tonB* | 271 | G A A T A T G A T T G C T A T T T G C |
| *exbB* | 506 | G A G A A C G A C T A T C A A T T C G |
| *exbC* | 1118 | A G C A A C G G C A A T C G G C C T C |
| *fepB* | -13 | G A A A A T G A G A A G C A T T A T T |
| *entC* | 7 | A T A A A T G A T A A T C A T T A T T |
| *fes* | 391 | G C A A A T G C A A A T A G T T A T C |
| *fes* | 415 | T A T T A T C A A T A T A T T T C T G |
| *fur* | 186 | T A T A A T G A T A C G C A T T A T C |
| *fecI* | 1170 | T G T A A T G A T A A C C A T T C T C |
| *fecI* | 1176 | G A T A A C C A T T C T C A T A T T A |
| *sltA* | 204 | G A A T A T G A T T A T C A T T T T C |
| *sodA* | 48 | G A T A A T C A T T T T C A A T A T C |
| Consensus | | G A T A A T G A T A A T C A T T A T C |
| *Serratia marcescens* | | |
| *shlB* | 320 | G A T T G T C A T A A T T T C C C C C |
| *sfuA* | 54 | T T T A A T A C G A A T C G T T T T C |
| *Vibrio cholerae* | | |
| *hly* | | A A T A A T A T G A A T A T C A G T A |
| *Corynebacterium diphtheriae* | | |
| *tox* | -56 | T A T A A T T A G G $N_9$C C T A A T T A T T |
| *Anacystis nidulans* | | |
| *isiA* | -51 | A C T T A TTG A C A A T T A T T G T A |
| *isiA* | -61 | C T T A A TAT C A A C T T A T T G A G |
| *Synechococcus* sp. | | |
| *irpA* | -212 | T A A A A A T G A $N_9$T C A T T T T T A |

a) Nucleotide numbering according to the publications listed in (32).

**Fig. 2.** Consensus sequences upstream of iron-regulated genes. In some cases there is more than one homologous sequence which may or may not play a regulatory role. Binding of the $Fe^{2+}$-Fur repressor has only been shown to the operator region upstream of *aer* and *cir*

## 3 Positive Selection Procedure for Iron Regulation Mutants

The experimentally possible replacement of $Fe^{2+}$ with $Mn^{2+}$ resulted in a convenient procedure to isolate repressor-negative mutants [7]. It was shown that more than half of the mutants resistant to high concentrations (10 mM) of $Mn^{2+}$ were mutated in the *fur* gene. Apparently, high concentrations of $Mn^{2+}$ stop growth because they repress the iron supply systems. Similarly, overproduction of the Fur protein inhibits growth (see below). Mutants in the *fur* gene were devoid of repression so that $Mn^{2+}$ was without effect. The rest of the resistant cells were presumably mutated in $Mn^{2+}$ transport resulting in a subtoxic intracellular $Mn^{2+}$ concentration.

## 4 Properties of Cells Carrying Mutations in a Plasmid-Encoded *fur* Gene

The physical state of active Fur protein (monomer, dimer, multimer) is unknown. Therefore, we employed the genetic in vivo technique of negative complementation to obtain insights into the active form of the Fur repressor. Insertion of an inactive polypeptide into a protein consisting of two or more identical polypeptides can result in a partially or fully inactive protein. We first constructed a Fur protein derivative in which 12 amino acids at the N-terminal end were replaced with 12 amino acids of the N-terminal end of β-galactosidase. A 500-base-pair *Hae*III/*Alu*I fragment of plasmid pMH15 (Fig. 3) was cloned into the *Hinc*II site of plasmid pUC19. The restriction enzyme *Hae*III cleaves in codon 12 of the *fur*

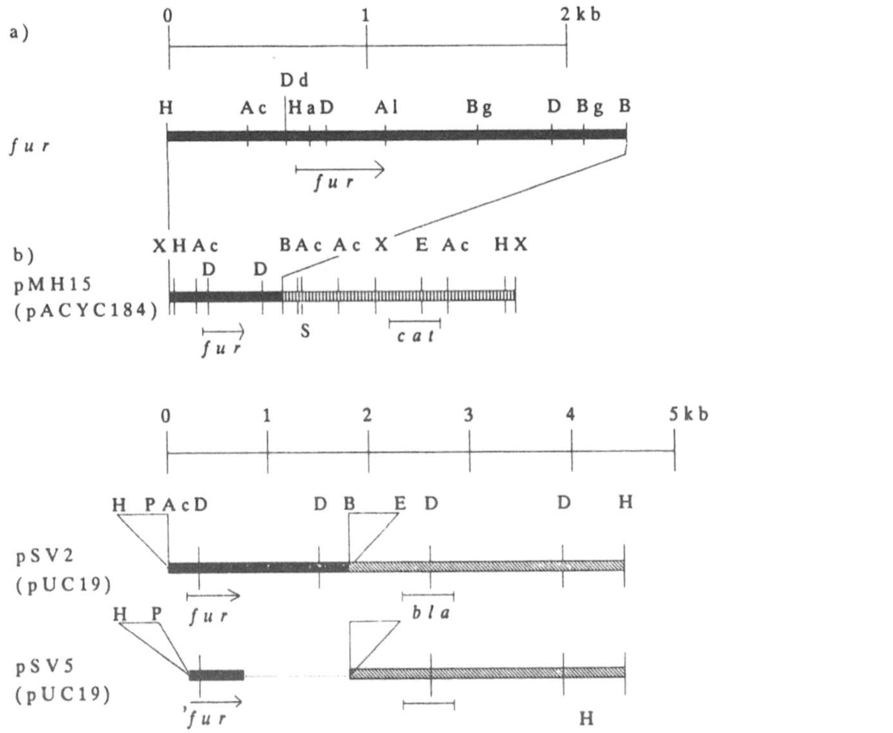


**Fig. 3. a.** The *fur* gene on plasmids; **b** pMH15, pSV2 and pSV5. The following restriction sites are indicated: Ac (*Acc*I), Al (*Alu*I), B (*Bam*HI), Bg (*Bgl*I), D (*Dra*I), Dd (*Dde*I), E (*Eco*RI), Ha (*Hae*III), H (HindII), S (*Sph*I), X (*Xho*II). The location of the *fur* gene (*fur*), the 5'truncated *fur* gene (*'fur*), the *cat* gene (chloramphenicol resistance), and the *bla* gene (ampicillin resistance) are indicated. The vectors used for cloning were pACYC184 and pUC19. The *triangles* indicate the multiple cloning sites of pUC19

gene while *Alu*I cleaves downstream of the transcriptional terminator of *fur*. Ligation into the *Hinc*II site results in an in-frame fusion with the *lacZ* gene. The 15 N-terminal amino acids of the Fur protein have the sequence M T D N N T A L K K A G L K V; the constructed fusion reads T M I T P S L H A C R S L K V.

Cells transformed with the constructed plasmid, termed pSV5 (Fig. 3), synthesized the fusion protein in sufficiently high amounts that it represented the strongest band on a polyacrylamide gel loaded with the soluble cell fraction (Fig. 4). Interestingly, much more fusion protein was synthesized than wild-type Fur protein (Fig. 4, lanes 1, 2) although a wild-type *fur* gene was cloned in the same vector (pUC19) as the mutated *fur* gene (Fig. 3). Wild-type *Fur* probably repressed its own synthesis, suggesting that the Fur consensus sequence upstream of the *fur* gene (Fig. 2) indeed functions as a Fur binding site. The apparent lack of autorepression shows that the Fur derivative is inactive. This conclusion is supported by the following observations. Overproduction of the wild-type repressor led to a slow growth rate, probably because repression of iron-regulated genes impaired the iron-dependent metabolism. In contrast, cells carrying pSV5 grew normally. In fact, pSV5 did not restore Fur repression of a *fur*⁻ mutant. The high expression of iron-regulated outer membrane proteins remained at this level after transformation of strain H1681 *fur*⁻ with plasmid pSV5 (Fig. 5, lane 3). Moreover, strain H1680 *fur*⁺ was partially converted to a Fur⁻ phenotype (lane 2, compare with lane 1 H1680 *fur*⁺, lane 4 H1681 *fur*⁻, and lane 5 H1681 *fur*⁻ transformed with plasmid pMH15 *fur*⁺). These results were supported by $\beta$-galactosidase activities of cells carrying a chromosomal *fhuF*::*lacZ* fusion. This operon fusion between the

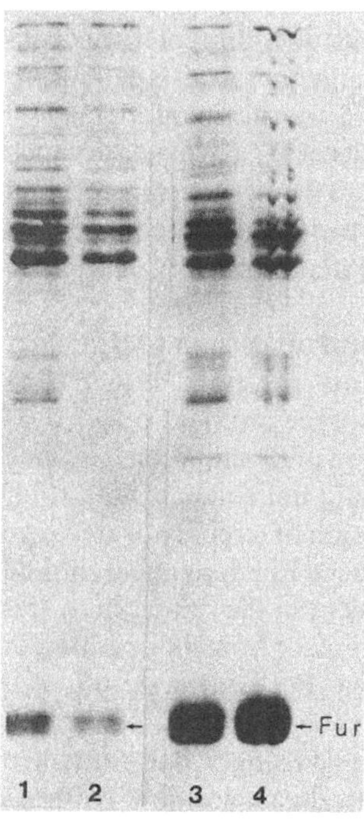


**Fig. 4.** Cytoplasmic proteins of *E. coli* H1681 *fur*⁻ transformed with plasmids pSV2 (lanes *1, 2*) and pSV5 (lanes *3, 4*). Cells were grown in the absence (lanes *1, 3*), and in the presence (lanes *2, 4*) of 0.3 mM IPTG (isopropyl-thiogalactoside). The proteins of the cytoplasmic fraction were separated on a 15% polyacrylamide gel in the presence of 0.1% sodium dodecyl sulfate (SDS-PAGE), and stained with Coomassie blue. *Fur* indicates the position of the Fur protein

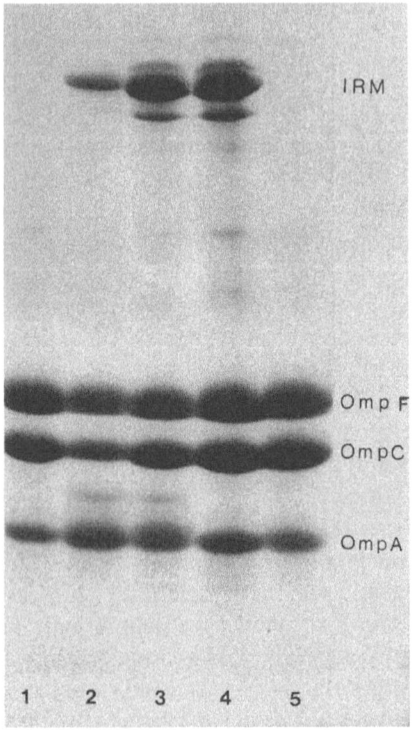


**Fig. 5.** Outer membrane proteins of *E. coli* H1680 *fur*$^+$ (lane *1*), transformed with plasmid pSV5 (lane *2*); of strain H1681 *fur*$^-$ transformed with pSV5 (lane *3*), untransformed (lane *4*); transformed with pMH15 (lane *5*). The cells were grown in iron-rich tryptone-yeast extract medium. Note the strong expression of the iron-regulated proteins (IRM) in the chromosal *fur*$^-$ strain transformed with the 5′ deleted *fur* derivative (lane *3*) in contrast to the *fur* wild-type transformant (lane *5*). The proteins were separated by SDS-PAGE and stained with Coomassie blue

iron-regulated *fhuF* gene and the *lacZ* gene lacking its own promoter was constructed using a phage λ p*lac* Mu53 derivative [26]. In the *fur*$^+$ wild-type 17 units of enzyme activity were measured, whereas the pSV5 transformant yielded 63 units. The chromosomal *fur*$^-$ mutant expressed 216 units of β-galactosidase which was reduced to 79 units in the pSV5 transformant, and to 9 units in a transformant carrying the wild-type *fur* gene on pSV2. These results show that the Fur fusion protein has retained a residual repressor activity, and that it partially eliminated wild-type Fur repressor activity.

Additional *fur* mutations were isolated by treatment of plasmid pMH15 *fur*$^+$ with hydroxylamine and selecting for transformants of strain H1680 *fur*$^+$ *fhuF*::*lacZ*, which showed a repressor-negative phenotype on MacConkey agar plates (red colonies). This approach avoided the isolation of *fur* mutations carrying large deletions, as the truncated Fur polypeptides could not compete with wild-type Fur repressor. It was assumed that for the expression of a repressor-negative phenotype in a chromosomal *fur*$^+$ strain, the overproduced Fur derivatives formed inactive heteromers with wild-type Fur repressor present in the same cell. A less likely alternative was the displacement of wild-type repressor from the operator, or inactivation of wild-type repressor by sequestering $Fe^{2+}$ through the Fur derivatives.

Three mutated plasmids were obtained. Two carried single, one contained three nucleotide substitutions in the *fur* gene (Table 1). β-galactosidase synthesis

**Table 1.** Properties of the negative complementing *fur* mutants[a]

| Protein designation | Plasmid | Codon change | Amino acid replacement | β-Galactosidase activity H1717 | SC11 |
|---|---|---|---|---|---|
| | – | – | – | 27 | 169 |
| Fur | pMH15 | – | – | 23 | 8 |
| Fur90 | pMH151 | CAC/TAC | His/Tyr | 113 | 135 |
| Fur51 | pMH152 | GGT/GAT | Gly/Asp | 181 | 154 |
| Fur11/12 | pMH154 | AAG/AAA | Lys/Lys | 85 | 122 |
| | | GCT/ACT | Ala/Thr | | |
| | | GGC/GAC | Gly/Asp | | |

[a] The numbers of Fur 90, 51, and 11/12 designate the mutated codons. Fur11/12 carries an additional mutation at codon 9 which, however, does not alter the Lys residue at this position. Strain H1717 is *fur*$^+$, SC11 is a *fur*$^-$ derivative thereof (*fur31*). The β-galactosidase activities of the iron-regulated *fhuF::lacZ* operon fusion are given in nM ortho-nitrophenyl β-D-galacto-pyranoside, hydrolyzed per min by $10^9$ cells.

was derepressed in a chromosomal *fur*$^+$ as well as in a *fur*$^-$ strain (Table 1). The latter shows that the Fur repressor derivatives were also inactive as a homomer. The degree of derepression differed among the three mutants. Plasmid pMH152 (Gly51Asp) showed the strongest derepression regarding synthesis of β-galactosidase and of iron-regulated outer membrane proteins (Fig. 6, compare lane 3 with lanes 2, 4, 5). In addition, cells carrying the mutated plasmids grew very slowly on succinate as a carbon source. This property is typical for *fur* mutants. Growth was not completely abolished, as was observed with *fur* insertion mutants, which agrees with the residual Fur repressor activity revealed by the level of β-galactosidase and outer membrane proteins.

Derepression by mutated *fur* genes on plasmids pMH151, 152 and 154 (plasmid pACYC184 derivatives) was also studied in cells carrying the wild-type *fur* gene on plasmid pBR322. No derepression was observed. The β-galactosidase values of a *fhuF::lacZ* fusion strain were 53, 49, and 72 units, which were close to the activity of cells carrying only the *fur* wild-type plasmid (33 units), or which were *fur*$^+$ on the chromosome (63 units). This indicates that the mutated Fur derivatives had to be present in excess over wild-type Fur in order to eliminate wild-type Fur repressor activity.

## 5 In-Vitro Activity of the Isolated Fur Derivatives

As can be expected for a regulatory protein, the cellular concentration of Fur repressor is low. In order to isolate the protein, the amount of Fur produced had to be strongly increased. This was achieved by cloning the wild-type and mutated *fur* genes downstream of the strong gene 10 promoter of phage T7 on plasmid pT7-6, and transcription by the very active T7 RNA polymerase. Synthesis of the polymerase was temperature regulated, which was important for cloning the wild-type *fur* gene, since increased amounts of the active repressor reduced the growth rate to more than half of cells expressing the chromosomal *fur* gene. After

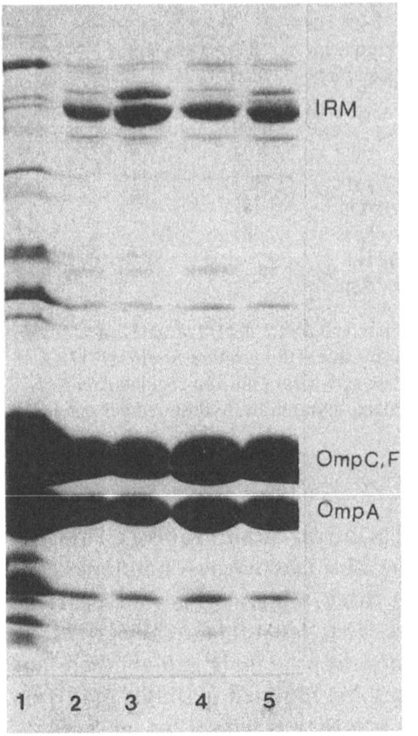


**Fig. 6.** Outer membrane proteins of *E. coli* H1717 *fur*⁺ (lane *1*) transformed with plasmids pMH151 (lane *2*); pMH152 (lane *3*); and pMH154 (lane *5*) encoding point mutations in the *fur* gene. Lane *4* contains the sample of an independently isolated plasmid mutation which was identical to that of pMH151 (lane 2). Cells were grown in iron-rich tyrptone yeast extract medium. Protein of the outer membrane fraction were separated by SDS-PAGE and stained with Coomassie blue. IRM, iron-regulated membrane proteins in the molecular weight range of 80 000 daltons. The positions of the major outer membrane proteins OmpC, OmpF, and OmpA are indicated. The gel has been purposely overloaded to demonstrate the IRM's

temperature induction (42 °C), the Fur protein formed a prominent band on a gel loaded with the cytoplasm (S-100 fraction) of cells carrying wild-type as well as mutated *fur* genes (Fig. 7, lanes 1, 3, 4). The Fur protein was purified by metal chelate affinity chromatography on Sepharose 6B, to which iminodiacetic acid groups were coupled by stable ether linkages through a long hydrophilic spacer (Pharmacia). The gel was charged with $Zn^{2+}$, to which the Fur protein was preferentially bound via its histidine residues. It was eluted by a buffer containing 20 mM histidine. A largely pure protein was obtained in a single step (Fig. 7, lane 5). The Fur derivatives were eluted at the same position as the wild-type Fur protein.

The isolated proteins were used to determine binding to the iron-regulated aerobactin promoter sequence on plasmid pRG132 [27]. Protection of the *Hinf*I cleavage site in the aerobactin promoter region by the Fur protein results in a 350-base pair fragment which is not cleaved into the 225 and 125 fragments. Indeed, prevention of *Hinf*I cleavage was observed at a Fur protein concentration of 100 nM and higher (Fig. 8, lanes 5–7), while the Fur51 derivative was inactive up to 6 μM (Fig. 8, lanes 8–12; Fig. 9, lane 5). The Fur11/12 and Fur90 derivatives also did not inhibit cleavage (data not shown). In vitro competition experiments between the Fur derivatives and wild-type Fur were also carried out to confirm the in-vivo negative complementation. The three derivatives prevented protection of the *Hinf*I site by Fur at a fivefold excess over Fur (Fig. 9, compare lanes 2, 3, 4, in

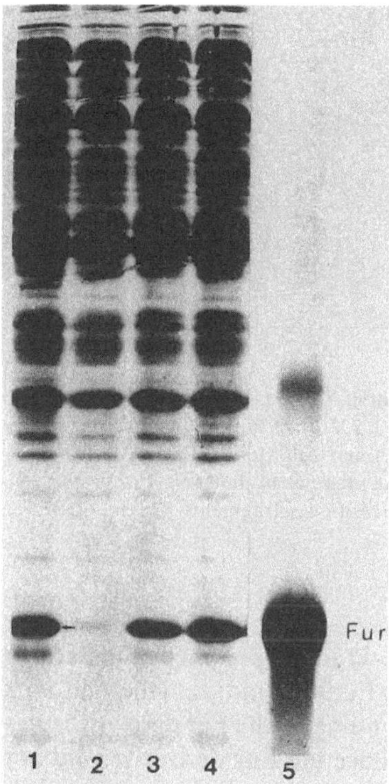


**Fig. 7.** Cytoplasmic fraction (S-100 supernatant) of *E. coli* W1576 transformed with plasmids expressing the wild-type *fur* gene (lane *1*); untransformed W1576 (lane *2*); W1576 transformed with *fur90* (lane *3*); *fur51* (lane *4*) synthesized under the transcriptional control of the gene 10 promoter and the RNA polymerase of phage T7. A purified Fur protein run on a separate gel has been added to the figure (lane *5*) to demonstrate the purity of the samples used for the DNA protection experiments. The gel was purposely overloaded to detect trace amounts of contaminating proteins (upper band)

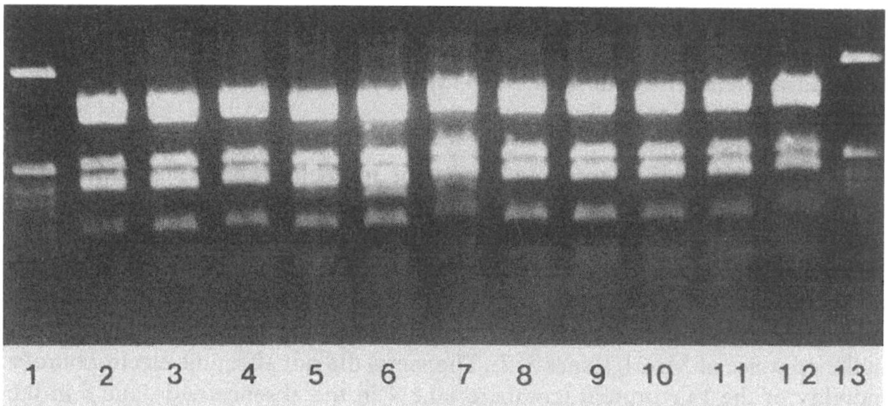


**Fig. 8.** Restriction fragments of plasmid pRG132 (27) cleaved with *HinfI* in the absence (lane *2*) and in the presence of increasing concentrations of isolated wild-type Fur protein (lanes *3* to *7*, 30, 90, 300, 900, 3000 nM, respectively), and of the mutated Fur51 protein (lanes *8* to *12*, 20, 60, 200, 600, 2000 nM). The *arrow* within the figure indicates the restriction fragments which is not cleaved in the presence of wild-type $Mn^{2+}$-Fur protein (lanes *5*, *6*, *7*). The samples contained 100 $\mu$M $MnCl_2$. Lanes *1* and *13* show the restriction fragments of *HinfI*-cleaved pBR322 used as standards

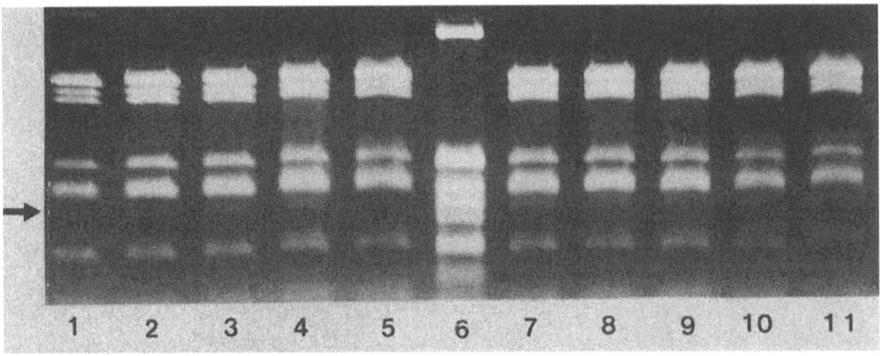


**Fig. 9.** Inhibition of wild-type Fur cleavage protection by Fur90. The experiment was done as in Fig. 8. The samples contained 0.4 μM (lane *2*) and 0.8 μM (lanes *3, 4, 7–11* of Fur protein; and *6* μM (lane 5), 0.5 μM (lane *7*), 1 μM (lane *8*), 2 μM (lane 9), 4 μM (lane *10*), and 6 μM (lane *11*) of Fur90. The *arrow* points to the restriction fragment which is not cleaved in the presence of the $Mn^{2+}$-Fur protein. No protein was added to the sample in lane *1*. Lane *6* contains the restriction fragments of plasmid pBR322 cleaved with *HinfI*

the presence of wild-type Fur alone,with lanes 7–11 in the presence of increasing amounts of additional Fur90; lane 5 contains only Fur90). Binding of the wild-type Fur protein to the promoter, and prevention of binding by the Fur derivatives was to the same extent $Mn^{2+}$-dependent (100μM). Since the Fur derivatives did not prevent *HinfI* cleavage at a 20-fold higher concentration than wild-type Fur, they do not inhibit wild-type Fur by competing for promoter binding nor for $Mn^{2+}$ binding. Rather, the data suggest that the derivatives inactivate wild-type Fur by forming inactive mixed oligomers.

## 6 Conformational Alterations of Wild-Type and Mutated Fur Repressor Proteins Upon $Mn^{2+}$ Binding

$Mn^{2+}$ converted the Fur protein to a repressor, implying a conformational change upon metal binding. We used trypsin sensitivity as a tool to examine the presumed structural transition as a result of $Mn^{2+}$ binding. Wild-type Fur protein (3 μg) in the absence of $MnCl_2$ was slowly degraded by 0.1 and 0.3 μg per ml trypsin during 1 h incubation at 37°C (Fig. 10A, lanes 1, 2). It required 1 μg/ml trypsin to yield a substantial degradation (lane 3). In contrast, degradation proceeded much faster in the presence of $MnCl_2$ (lanes 5–7). The metal did not affect the electrophoretic mobility of the Fur protein (compare lane 4 in the absence with lane 8 in the presence of $MnCl_2$), nor did it activate trypsin since the rate of bovine serum albumin degradation was independent of $MnCl_2$. Mutant Fur51 showed similar properties, indicating that it could bind $Mn^{2+}$, and reacted by a conformational change as wild-type Fur.

Trypsin sensitivity of Fur90 was greater in the absence than in the presence of $MnCl_2$ (Fig. 10B, cf. lanes 1–3 with lanes 5–7). Apparently, Fur90 was able to bind

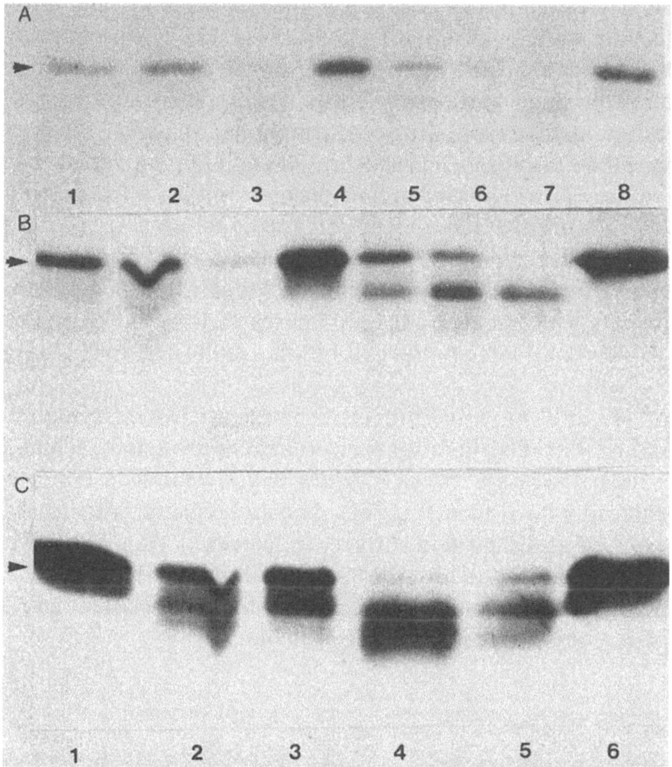


**Fig. 10. A** Cleavage of isolated wild-type Fur protein; **B** of Fur90; **C** of Fur11/12 by trypsin. The samples 5 to 8 (**A**, **B**), and 3,5, 6 (**C**) contained 5 mM $MnCl_2$. The concentrations of trypsin were : 0.1 µg/ml (**A**, **B**, lanes *1*, *5*; **C**, lanes *2*, *3*), 0.3 µg/ml (**A**, **B**, lanes *2*, *6*; C, lanes *4*, *5*), and 1 µg/ml (**A**, **B**, lanes *3*, *7*). No trypsin was added to the samples in lane *4* (**A**, **B**) and *1*, *6* (**C**). **A**, **B** were stained with Coomassie blue, **C** with silver nitrate. The *arrowheads* point to the undegraded Fur proteins

$Mn^{2+}$ but the His/Tyr substitution affected the conformation of the protein such that the 16 kDa degradation product was largely resistant to further degradation.

Degradation of Fur 11/12 by trypsin was unaffected by $MnCl_2$ (Fig. 10C, cf. lanes 2,4 in the absence, with 3,5 in the presence of $MnCl_2$). This protein was either incapable of binding $Mn^{2+}$, or to react by a conformational change.

## 7 Secondary Structure Analysis of the Fur Repressor and the Inactive Fur Derivatives

The Fur protein and its mutated derivatives were analyzed by version 88 of the HYCON program (W. Tröger, unpubl.) which combines secondary structure predictions for hydrophilicity, hydropathy, acrophily, antigenicity, and flexibility of a polypeptide. The 12 amino acids of the N-terminal end of β-galactosidase, which replaced the 12 amino acids at the N-terminus of the Fur protein, conferred to the resulting Fur derivative in this region a higher hydropathy, and a lower hydrophilicity, acrophily, antigenicity, and flexibility (Fig. 11). The probability of forming an $\alpha$-helix was decreased and of forming a $\beta$-pleated sheet was increased. Apparently, the relatively small sequence alterations with regard to the physical properties of the amino acids resulted in conformational changes which were sufficient to inactivate the Fur repressor. The Ala/Thr and Gly/Asp replacements of Fur11/12 in the same region increased mainly the hydrophilicity and flexibility. The residual Fur11/12 repressor activity support the conclusion regarding the importance of the N-terminal region for repressor activity. The Gly/Asp substitution in Fur51 primarily increased the hydrophilicity of that region (Fig. 11). Since the change in conformation is so small, we suggest this region to be part of the active site, in that the negatively charged Asp prevents binding to the negatively charged DNA backbone.

The exchange of His by Tyr in Fur 90 affects an amino acid which is frequently involved in iron binding. For this alteration a profound conformational change is predicted, in that a helical region between residues 86 and 96 in wild-type Fur is interrupted and replaced by a $\beta$-turn (Fig. 11). The antigenicity, which usually concerns regions exposed at the protein surface, is decreased. As will be shown below, the region between residues 83 and 92 shows homologies to helix 2 of DNA-binding proteins with a helix-turn-helix motif, which is consistent with the location of this region at the surface of the Fur protein.

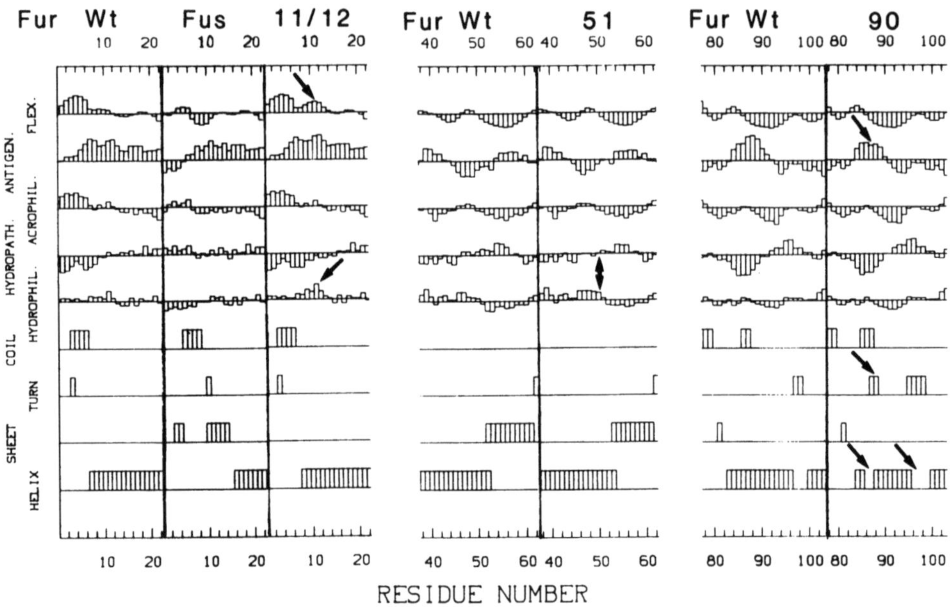


**Fig. 11.** Secondary structure prediction of wild-type Fur protein (Fur Wt), fusion protein (Fus), Fur 11/12 (11/12), Fur51 (51), and Fur90 (90) in the mutated regions. The *arrows* point to the most prominent changes

## 8 Functional Domains of the Fur Protein

Based on comparisons with other DNA binding proteins, we have previously proposed that the region between residues 114 and 129 may form a helix which binds to the DNA operator region [19]. The same region is identified by a statistical method which is based on known and likely DNA-binding proteins of the Cro type. They are used to derive a master set of aligned sequenes which is employed to count the number of occurrences of each amino acid at each position of the protein of interest, in this case of the Fur repressor. Scores of at least 1400 represent a significant chance of a protein being Cro-like [28]. The sequence of Fur with the highest score, 1029, extends from residue 111–120. The second largest value, 458, is obtained for residues 83–92. The 2 added values (1486) strongly point to two helical DNA-binding regions which, however, differ from the helix-turn-helix motif, in that the two helices are arranged some distance from each other, and in an inverted order. This is not unexpected in the light of the aporepressor nature of the Fur protein which is converted to a repressor by the corepressor $Fe^{2+}$. The conformational change induced by binding the metal ion could bring together the two helices so that they fit into the grove of the DNA helix. The Fur protein contains the sequences Cys-$X_4$-Cys-$X_4$-His-X-His and $His_3$-X-His-$X_2$-Cys-$X_2$-Cys, where X is an arbitrary amino acid. Similar sequences are frequently found in nucleic acid-binding proteins (for example, zink finger proteins), and in redox enzymes, (see, for example, [29, 30]), for which metal ion binding has either been demonstrated or suggested. However, the Fur regions are not identical to any one of these sequences. It is likely that the His, Cys, and also Tyr residues are involved in $Fe^{2+}$ binding and in the structural transitions, which increase the affinity for Fur binding sites in the operator regions of iron-regulated genes, although this has still to be demonstrated directly.

## 9 Conclusions

The Fur protein is converted to a repressor upon binding of $Fe^{2+}$. The active repressor has an oligomeric structure. The Fur repressor not only regulates iron supply systems but rather serves to adapt bacteria to environments which usually are low in available iron. The Fur-type iron regulation of *E. coli* occurs in many bacteria, including gram-positives.

*Acknowledgments.* We thank Tony Focareta for valuable comments on the manuscript. This work was supported by the Deutsche Forschungsgemeinschaft (SFB 323) and the Fonds der Chemischen Industrie.

## References

1. Archibald, F. (1983) *Lactobacillus plantarum*, an organism not requiring iron. FEMS Microbiol. Lett. 19:29–32
2. Makemson, J.C. & Hastings, J.W. (1982) Iron represses bioluminescence and affects catabolite repression of luminescence in *Vibrio harveyi*. Current Microbiol. 7:181–186

3. McCarter, L. & Silverman, M. (1989) Iron regulation of swarmer cell differentiation of *Vibrio parahaemolyticus*. J. Bacteriol. 171:731–736
4. Niederhoffer, E.C., Naranjo, C.M. & Fee, J.A. (1989) Relationship of the superoxide dismutase genes, *sodA* and *sodB*, to the iron uptake (*fur*) regulon in *Escherichia coli* K-12. In: Metal ion homeostasis: molecular biology and chemistry. Liss, New York, pp. 149–158
5. Laudenbach, D.E. & Straus, N.A. (1988) Characterization of a cyanobacterial iron stress-induced gene similar to *psbC*. J. Bacteriol. 170:5018–5026
6. Henderson, N., Austin, S. & Dixon, R.A. (1989) Role of metal ions in negative regulation of nitrogen fixation by the *nifL* gene product from *Klebsiella pneumoniae*. Mol. Gen. Genet. 216:484–491
7. Hantke, K. (1987) Selection procedure for deregulated iron transport mutants (*fur*) in *Escherichia coli* K12: *fur* not only affects iron metabolism. Mol. Gen. Genet. 210:135–139
8. Braun, V. & Winkelmann, G. (1987) Microbial iron transport, structure and function of siderophores. Progress Clin. Biochem. Med. 5:67–99
9. Braun, V., Hancock, R.E.W., Hantke, K. & Hartmann, A. (1976) Functional organization of the outer membrane of *Escherichia coli*: phage and colicin receptors as components of iron uptake systems. J. Suparamol. Struct. 5:37–58
10. Bennett, R.L. & Rothfield, L.J. (1976) Genetic and physiological regulation of intrinsic proteins of the outer membrane of *Salmonella typhimurium*. J. Bacteriol. 127:498–504
11. Casadaban, M.J. & Cohen, S.M. (1979) Lactose genes fused to exogenous promoters in one step using a Mu-*lac* bacteriophage: in vivo probe for transcriptional control sequences. Proc. Natl. Acad. Sci. USA 76:4530–4533
12. Hantke, K. (1981) Regulation of ferric iron transport in *Escherichia coli* K12: isolation of a constitutive mutant. Mol. Gen. Genet. 182:288–292
13. Braun, V. & Burkhardt, R. (1982) Regulation of the ColV plasmid-determined iron (III)-aerobactin transport system in *Escherichia coli*. J. Bacteriol. 152:223–231
14. Hantke, K. (1983) Identification of an iron uptake system specific for coprogen and rhodotorulic acid in *Escherichia coli* K12. Mol. Gen. Genet. 191:301–306
15. Hantke, K. & Zimmermann, L. (1981) The importance of the *exbB* gene for vitamin B12 and ferric iron transport. FEMS Microbiol. Lett. 12:31–35
16. Braun, V., Gross, R., Köster, W. & Zimmermann, L. (1983) Plasmid and chromosomal mutants in the iron(III)-aerobactin transport system of *Escherichia coli*. Use of streptonigrin for selection. Mol. Gen. Genet. 192:131–139
17. Zimmermann, L., Hantke, K. & Braun, V. (1984) Exogenous induction of the iron dicitrate transport system of *Escherichia coli* K12. J. Bacteriol. 159:271–277
18. Braun, V. (1985) The iron-transport system of *Escherichia coli*. In: Martonosi, A.N. (ed.) The enzymes of biological membranes, vol 3. Plenum, New York, pp. 617–652
19. Hantke, K. (1984) Cloning of the repressor protein gene of iron-regulated systems in *Escherichia coli K12*. Mol. Gen. Genet. 197:337–341
20. Schäffer, S., Hantke, K. & Braun, V. (1985) Nucleotide sequence of the iron regulatory gene *fur*. Mol. Gen. Genet. 200:110–113
21. Wee, S., Neilands, J.B., Bittner, M.L., Hemming, B.C., Haymore, B.L. & Seetharam, R. (1988) Expression, isolation and properties of Fur (ferric uptake regulation) protein of *Escherichia coli* K12. Biol. Metals 1:62–68
22. de Lorenzo, V., Giovannini, G., Herrero, M. & Neilands, J.B. (1988) Metal ion regulation of gene expression. Fur repressor-operator interaction at the promoter region of the aerobactin system of pColV-K30. J. Mol. Biol. 20:875–884
23. Bagg, A. & Neilands, J.B. (1987) Ferric uptake regulation protein acts as a repressor, employing iron(II) as a cofactor to bind the operator of an iron transport operon in *Escherichia coli*. Biochem. 26:5471–5477
24. Griggs, D.W. & Konisky, J. (1989) Mechanism for iron-regulated transcription of the *Escherichia coli cir* gene: metal-dependent binding of Fur protein to the promoters. J. Bacteriol. 171:1048–1054
25. Calderwood, S.B. & Mekalanos, J.J. (1988) Confirmation of the Fur operator site by insertion of a synthetic oligonucleotide into an operon fusion plasmid. J. Bacteriol. 170:1015–1017
26. Bremer, E., Silhavy, T.J. & Weinstock, G.M. (1985) Transposable lambdaplacMu bacteriophages for creating *lacZ* operon fusions and kanamycin resistance insertions in *Escherichia coli*. J. Bacteriol. 162:1092–1099

27. Gross, R., Engelbrecht, F. & Braun, V. (1985) Identification of the genes and their polypeptide products responsible for aerobactin by pColV plasmids. Mol. Gen. Genet. 201:204-212
28. Dodd, I.B. & Egan, J.B. (1987) Systematic method for the detection of potential Lambda Cro-like DNA-binding regions in proteins. J. Mol. Biol. 194:557-564
29. Fischer, H.-M., Bruderer, T. & Hennecke, H. (1988) Essential and non-essential domains in the *Bradyrhizobium japonicum* NifA protein: identification of indispensable cysteine residues potentially involved in redox reactivity and/or metal binding. Nucl. Acids Res. 16:2207-2224
30. Trageser, M. & Unden, G. (1989) Role of cysteine residues and of metal ions in the regulatory functioning of Fnr, the transcriptional regulator of anaerobic respiration in *Escherichia coli*. Mol. Microbiol. 3:593-599
31. Chehade, H. & Braun, V. (1988) Iron-regulated synthesis and uptake of colicin V. FEMS Microbiol. Lett. 52:177-182
32. Braun, V. & Hantke, K. (1991) Genetics of bacterial iron transport. In: Winkelmann, G. (ed.) Handbook of microbial iron chelates (siderophores). CRC Press, Boca Raton (in press)

# Subject Index

MIX
Papier aus verantwortungsvollen Quellen
Paper from responsible sources
FSC® C105338

If you have any concerns about our products,
you can contact us on
**ProductSafety@springernature.com**

In case Publisher is established outside the EU,
the EU authorized representative is:
**Springer Nature Customer Service Center GmbH**
**Europaplatz 3, 69115 Heidelberg, Germany**

Printed by Libri Plureos GmbH
in Hamburg, Germany